DEMOCRACY AND PEACE

DEMOCRACY AND PEACE

By

Dr. M. Lakshmi Narasaiah

M.A., Ph.D.

Professor of Economics,
Coordinator, Department of M.B.A.
Sri Krishnadevaraya University Post-graduate Centre,
Kurnool–518 002
Andhra Pradesh (India)

DISCOVERY PUBLISHING HOUSE
NEW DELHI

Reprinted - 2020

First Published - 2005

ISBN: 978-81-7141-930-2

Democracy and Peace

Published by:

DISCOVERY PUBLISHING HOUSE PVT. LTD.
4383/4B, Ansari Road, Darya Ganj
New Delhi-110 002 (India)
Phone: +91-11-23279245, 23253475; 43596065
E-mail: discoverybooksindia@gmail.com
discoverypublishinghouse@gmail.com
web: www.discoverypublishinggroup.com

Printed at:
Infinity Imaging Systems
Delhi

Preface

Democracy assistance and poverty reduction are rightly becoming two focal and related-issues for development assistance. Increasingly, many organisations, including intergovernmental, national and civil society, are focusing their work on these two area. Futhermore, the relationship between these two issues is complex and ever changing. There is thus a need to develop methodologies of linking democracy assistance and poverty reduction at both the policy and programme levels. International IDEA (Institute for Democracy and Electoral Assistance) in cooperation with the World Bank and the United Nations Development Programme, is developing concrete strategies that address these two objectives in a mutually reinforcing way. Through an overall situation analysis followed by regional meetings in sub-Saharan Africa, South Asia, Latin America, the Caucasus and the Arab region, the Institute has marshalled evidence of some of the key problems that affect democracy consolidation and poverty reduction in these countries:

- Corruption and its undermining effect on popular confidence in public institutions.
- Continuing economic instability coupled with the lack of strategies for addressing the twin challenges of poverty and increasing popular participation in its alleviation.
- The extremely limited nature of citizen's influence on overall policy and decision-making processes despite the spread of formal democratic institutions.
- A trend in many post-communist states towards viewing growing poverty as a direct consequence of a transition to democracy.

In short, the evidence is not very encouraging for the prospects for democracy consolidation and poverty reduction. The critical step International IDEA advocates is the development of an approach that not only seeks to put democracy assistance and poverty reduction on top of the development assistance agenda, but also to encourage all involved to treat them as twin elements of an integrated programme of action.

Through a focus on accountable governance, promotion and protection of citizenship and rights and increased popular participation, International IDEA believes that both democracy and poverty reduction can be addressed simultaneously. Policy recommendations are being developed and will be shared in the course of this year with governments, international organisations and civil society bodies.

International IDEA believes that democracy promotion can be used as a tool for fulfilling a variety of objectives. Democracy matters because it protects human rights and preserves human dignity. But democracy also matters because it helps to address some of the most critical challenges facing states today: peace, development, economic growth and stability.

Democracy does not guarantee any one of these, but increasingly it seems to be a precondition for them in the long term. Thus, advocating democracy goes beyond being of a moral issue; it becomes *fundamental* to advancing the well-being people and the stability of states. International IDEA will continue to explore the link between democracy and the major issues facing society today—and continue to argue the case for democracy.

Dr. M. Lakshmi Narasaiah

Contents

1

Democracy and the Market Economy

Today the idea of democracy is triumphant; the model is in principle embraced in most countries the world over. You may say that the very word democracy has been hailed and misused earlier in history. The most repressing and totalitarian regimes have tried to mask themselves as "real" or "peoples" democracies. What has happened, however, is a historical demasking of these false pretences.

What exactly do we mean by democracy? There is now a general agreement that democracy cannot be defined by purpose or policy or levels of mass mobilisation. It must be defined as a political system where different parties or individuals compete for power through regular free elections where all adult citizens have a vote. Moreover, a democracy must uphold certain basic human rights and well-defined freedoms which make the political process possible, and respect the opinion and integrity of the individual. No other definitions hold, and we should be careful when we talk about "real" democracy versus "formal" democracy. A society, which in real life upholds the constitutional or formal democratic principles and which in practice applies the rights these principles imply, is by definition a democracy. A society with a beautiful-sounding constitution but where none or few of these rights are respected is certainly not a democracy.

Democratic Government no Guarantee for Equality

It is important to understand that democratic government does not necessarily mean good government in the sense that those in power make wise or well-considered decisions. Nor does it mean that conflicts inherent in the society are reduced to a minimum. Demands for democracy, social justice and a better life have historically gone hand in hand, but this does not mean that the establishment of a democratic system actually does lead to an improvement in social conditions or equality. It is also quite clear that some societies have a sort of outer shell of democracy but in reality, exclude large groups of people from having any political influence whatsoever. The actual differences in living conditions are so enormous and so entrenched that these people have no confidence at all in the political system even if it is democratic according to the definition. In these cases—for example in some Latin American countries—one can talk of a "masked hegemony with competing elites" where the outcome of struggles for power has little relevance for the masses. It is a sort of social and political half-authoritarian system—but disguised as a democracy—where the military often have a significant influence.

In the rhetoric of the day the term market economy and democracy are used as if they were synonymous or at least naturally emerging at the same time. But this is wrong—or at least misleading. When the market economy or capitalism finally established itself in the 1800s and came to characterize modern industrial civilisation, democracy was at best in its infancy. In fact one could argue that democracy grew out of the contradictions and social dynamism inherent in the market economy of the capitalistic system. In this century we have a long list of terrifying and repressive regimes, which have nevertheless upheld the virtues of a market economy. That some of these regimes have for ideological and security reason been hailed as bastions against communism, and also dignified members of the so-called free world does not transform them into democracies. In this company it is perhaps unnecessary to remind ourselves that the colonial system was assuredly not democratic, but was certainly based on capitalistic or market economic principles. It is the sad but irrefutable historical coupling between Western democracy, colonialism, and the plundering of resources in the name of the market economy which for understandable reasons meant that many of the leaders of national liberation movements looked for other models for the development of their young nations. In this connection it can be worth remembering what Nelson Mandela said soon after his release from 26 years of prison in the market economic but racist state of South Africa. "When we in ANC during 40 years struggled for democracy we were put in prison by the same people who are now telling us how we should behave to promote the democracy we have been rejected by all these year".

While we can see that a market economy does not automatically lead to democracy, a functioning democracy—as we have defined it—does seem to require some form of free economic system.

Democracy and Economic Freedom

Theoretically, it is conceivable that a political democracy could be combined with an economy totally controlled by the government—but experience has shown this to be very difficult. One could even argue that it is by definition impossible since democracy implies a certain freedom of economic choice and independent economic actors. A functioning democratic system presupposes what is now often referred to as a civil society—in practice, independent institutions, companies, organisations, the media etc., regulated by law but not subject to or controlled by those in power.

We must also see clearly that there are no unambiguous relations between economic growth, development and democracy. Democratic governments are neither very successful when it comes to structural reforms which may be to the disadvantage of important interest in the society, nor when it comes to welfare. The developing countries which have achieved the greatest success economically and socially over the last 20 years are the East Asian countries—which all have had various kinds of more or less authoritarian systems.

However, that does not mean that you can use these countries as models for the rest of the world. There is no globally valid link between an authoritarian form of regime and economic development, not even when development is defined only in terms of autocentric growth. Many social scientist—have tried to find some systematic connection between what we call development or modernisation on the one hand, and the political system on the other—but all have failed.

It is also obvious that one of several pre-requisites for economic growth and development is legitimate and reasonably well functioning government and governance. If the free market is to be a motor for development and improved welfare, and not just a meeting place for robber barons, the mafia and speculators, you must have a regulating and supportive state. If economic history teaches us anything, it is just this. Consider the astounding development in Germany after the war, or in Japan and the other East Asian countries some years later. There are many differences, but what they have in common is a well-functioning government apparatus with a long tradition.

Today we find ourselves in a historical situation where a large number of countries in the former communist states of Europe, in Africa, Asia and Latin America are at one and the same time trying to establish a new democratic system and new economic mechanisms. The situation is unique, and the intrinsic problems are unprecedented. Democracy as an idea has triumphed but in its practice it is in profound trouble. It is no exaggeration to talk of the crisis of democracy.

The former communist countries are certainly in crisis. As by-product of the past regimes, there is an intensive suspicion of the political institutions, of the state and the parties-and in this way also the legitimacy of democracy and the ability of the politicians to deal with the fundamental problems of society has been undermined. The lack of a democratic tradition is not overcome from one day to the next.

Many of the developing countries have similar difficulties. The introduction of a multiparty system does not in itself mean that one can manage the conflicts and social problems in a democratic way.

Countries in Transition

Both in the East and the South countries are trying, at one and the same time, to change the political and economic system.

When the whole society is convulsed by economic changes, and where peoples' living conditions fundamentally change, it is not easy to develop and maintain a political system based on compromise and respect, including respect for minorities.

As in previous history the deep crises of legitimacy and general frustration feed national and ethnical conflicts. These conflicts establish themselves in societies where the authoritarian system, economic crises and the break down of traditional values rob people of any kind of kinship other than ethnical.

We cannot avoid seeing disturbing signs of this crisis of democracy also in the so-called "established democracies" of the rich countries.

It is obvious that the state of democracy varies from country to country as do the reasons for a feeling of dejection. But there are some similarities too.

The continuing and noticeable internationalisation limits the national freedom of political choice, available alternatives, and makes it more difficult for people to see the connection between "politics" and their actual living conditions. The governments are restrained by international economic events. The reaction of the stock exchange may be more important than that of the voters. The election results influence the stock exchange prices—but is it perhaps not also so that the stock exchanges, indirectly, also influence the election results? People feel themselves to be the victims of major economic changes, but no one seems to be responsible and they themselves feel they have little chance of influencing the outcome. The absence of clearly identifiable alternatives between the larger political parties provides opportunities for the populists and the extremists.

There is indeed reason to reflect on the lessons of the history of our turbulent and cruel century.

Priority for Growth

There is today much concern about the lack of resources for such urgent needs as the reconstruction of the East, a concerted attack on poverty and human development in the poorest countries, and environmental investments of all kinds.

If the growth of world output returns to the levels of the 1980s, total output would grow by about one trillion dollar a year. There is, infact, no other way to resolve the economic and political crises multiplying in the world community than to give priority to the restoration of growth.

We are certainly not at the end of history as someone has argued. We are rather at a dramatic turning point, a moment of many possibilities and many dangers. What we do now, for a few years ahead, may direct the future for several decades—like the dramatic and fateful years immediately after the second world war. All nations, all governments, have a responsibility. The rich world has a special responsibility, not just moral because of its enormous economic and political power.

2

Democracy and Poverty:
Are they Interlinked?

Democracy assistance and poverty reduction are rightly becoming two focal–and related—issues for development assistance. Increasingly, many organisations, including intergovernmental, national and civil society, are focusing their work on these two areas. Futhermore, the relationship between these two issues is complex and ever changing. There is thus a need to develop methodologies of linking democracy assistance and poverty reduction at both the policy and programme levels. International IDEA (Institute for Democracy and Electoral Assistance) in cooperation with the World Bank and the United Nations Development Programme, is developing concrete strategies that address these two objectives in a mutually reinforcing way. Through an overall situation analysis followed by regional meetings in sub-Saharan Africa, South Asia, Latin America, the Caucasus and the Arab region, the Institute has marshalled evidence of some of the key problems that affect democracy consolidation and poverty reduction in these countries:

- Corruption and its undermining effect on popular confidence in public institutions.
- Continuing economic instability coupled with the lack of strategies for addressing the twin challenges of poverty and increasing popular participation in its alleviation.
- The extremely limited nature of citizen's influence on overall policy and decision-making processes despite the spread of formal democratic institutions.

- A trend in many post-communist states towards viewing growing poverty as a direct consequence of a transition to democracy.

In short, the evidence is not very encouraging for the prospects for democracy consolidation and poverty reduction. The critical step International IDEA advocates is the development of an approach that not only seeks to put democracy assistance and poverty reduction on top of the development assistance agenda, but also to encourage all involved to treat them as twin elements of an integrated programme of action.

Through a focus on accountable governance, promotion and protection of citizenship and rights and increased popular participation, International IDEA believes that both democracy and poverty reduction can be addressed simultaneously. Policy recommendations are being developed and will be shared in the course of this year with governments, international organisations and civil society bodies.

International IDEA believes that democracy promotion can be used as a tool for fulfilling a variety of objectives. Democracy matters because it protects human right and preserves human dignity. But democracy also matters because it helps to address some of the most critical challenges facing states today: peace, development, economic growth and stability.

Democracy does not guarantee any one of these, but increasingly it seems to be a precondition for them in the long term. Thus, advocating democracy goes beyond being a moral issue; it becomes *fundamental* to advancing the well-being of people and the stability of states. International IDEA will continue to explore the link between democracy and the major issues facing society today—and continue to argue the case for democracy.

3

City Politics: *A Voice for the Poor*

By 2020 the world's urban population will rise by almost 1.5 billion. Cities and towns house a growing proportion of poor people, partly because of the increased share of urban population of the total but also because economic recession and adjustment policies often hit poorer urban residents the hardest. Cities are associated with economic growth and wealth generation and yet inequality is high. Poor people generally live in substandard conditions, may not benefit from job creation, and suffer high levels of pollution, crime and violence.

How can city Governments cope with the challenges of population growth and increased global economic competition, and meet the needs of poor residents/is urban Governance responsive to the needs of the poor? Are the agencies responsible for city Government, especially the municipalities, addressing poor people's needs? Are NGOs and people's organisations playing a greater role in service delivery? Or is their role one of advocacy and lobbying? If so, how do they relate to the formal political system? Can Governments fulfil their responsibilities, including poverty reduction? How can the well being of poor urban Governance institutions priorities their needs? In assessing the responsiveness of city Government to poor people, three key questions are addressed:

How can the Poor Influence the Agenda of the Institutions of Urban Governance?

The influence of poor residents on decision making is controlled, in part, by the formal political system. Democratisation gives people a voice. However, this vote means more when elected representative depend on the political support of poor people- where they are a majority, or are well organised, or where there is a ward-based system. If poor people are organised enough, to articulate their needs and demand a fair share of urban resources. NGOs can help poor groups organise better and provide support for networking.

Where poor people are not organised it does not mean they are politically powerless. Poor people in this situation, however, are prey to the disadvantages of patronage and unlikely to be included in formal consultative processes. For an electoral system to be truly responsive, specific mechanisms and channels, such as consultative and participatory processes at city an sub-city levels, are needed to complement representative democracy. Athough, these channels do not necessarily include the poorest or make a marked difference to resources allocation, pro-poor decisions are unlikely without them.

How Can Cities Finance their Activities and Reduce Poverty?

Democratisation has not, in many countries brought allocation of financial resources or the revenue-raising capacity for local governments to fulfil their responsibilities. The responsiveness of city governments to poor people's needs thus depends, on whose voices are heard in the arenas of political decision making. Responsiveness also depends on how available financial resources are allocated and how the programmes they finance are designed. There is scope, for city governments to increase property and business revenues, and to borrow for capital investment. Whether increased financial resources benefit poor people depends on how the demands of external investors and creditors are reconciled with the demands of poor residents; the willingness of politicians and officials to address the distributive implications of existing and planned spending; and efficient transparent financial management. If funds are made available to sub-city levels of government or if expenditure can be influenced by ward councilors, the funds might then be used to meet the priorities of poor residents.

What are the Necessities of Urban Living and how can Access to them be Ensured?

An Adequate Income: Work opportunities should be the top priority. City governments can, however, support the urban economy in general and the economic activities of the poor in particular. Firstly they can ensure that the basic services are efficiently provided. Secondly city governments can refrain from activities that destroy the assets and livelihoods of the poor, especially eviction of informal settlements and micro-enterprises. Savings and credit schemes can be more appropriately organised at a community level and supported by NGOs.

Land ownership is a common aspiration for poor households. A home with secure tenure (not necessarily title) provides security, an appreciating asset, access to services, and a base for economic activities. Increasing the opportunities for poor households to gain access to a well-located plot of land is an important component of any poverty reduction strategy. Many never fulfil their dream and the needs of those who cannot, or do not wish to become home owners should not be neglected, however.

Local government is potentially more responsive to poor residents than are central government agencies, although this depends on the balance of political power and bureaucratic perceptions. The limited ability of the public sector to secure benefits for the poor from public-private partnerships in land development, suggest that more informal arrangements and the involvement of CSOs may be better ways forward.

Environmental Services: Land alone will not reduce poverty but must be linked to a healthy living environment—a package of appropriate and affordable environmental services, such as public transport, water and sanitation, solid waste collection, and energy for cooking and lighting. Rather than discussing appropriate standards, detailed issues of financing and affordability or how continued provision can be assured for each of these services, the research focused on how far decision making channels, mechanisms and partnership arrangements ensure that providers are responsive to the needs and priorities of poor residents.

Collaborative planning and decision making arrangements are one promising alternative, despite, the current shortcomings of participatory Budgeting. For responsiveness to the poor to be built in to such processes, local bureaucrats need to change their attitudes and working practices. Is it possible and acceptable for poor people to have to rely on their own resources their households and networks—resources that are very limited? Informal networks and links can, however, provide mutual support and access to politicians and bureaucrats, community associations thought not always present, inclusive or transparent, can play an important role in articulating poor residents views and in organising self-help activities. There is scope for formal representative community organisations, for informal links between peoples' organsiations and the power structures, and for networking between people's groups. NGOs can play an important role in developing the capacity of community organisations and in facilitating networking. Where NGOs play a role in service delivery. However, there is a danger that the resulting close relationship with local government detracts from their ability to empower poor people and challenge inappropriate policies. City governments, it is clear, cannot cope with the challenges of population and economic growth and respond to the needs of poor people alone. Only in alliance with other actors is there some hope that poverty can be overcome. For CSOs, many of which were forged during struggles for democratisation, this implies moving beyond confrontation to engagement. To form alliances between CSOs and city governments that put the interests of the poor first, poor people must be able to exercise their political rights.

4

A Nuclear Weapon Free World:

That Dream Must Become Reality

Until recently the likehood of achieving a world without nuclear weapons was very, very small. But we are now living in a different world, and in this new configuration, what was a utopian dream yesterday can be the subject of serious discussions today, and put into practice tomorrow.

It was by a quirk of history that the conception of the atom bomb coincided with the start of the Second World War. The main motivation for the scientists who initiated the work on the atom bomb was that the bomb should not be used. Our argument was that we needed the bomb in order to deter Hitler from using his bomb against us. But as it happened our bombs were used, they were used as soon as they were made, and they were used against civilian populations. The bombs on Hiroshima and Nagasaki have brought the Second World War to a rapid end. But they also had another, effect, namely they demonstrated to the Soviet Union the newly acquired, overwhelming power of the United States. From the very beginning nuclear weapons were seen as a major tool in the ideological struggle between the United States and the Soviet Union.

With the end of the Cold War, and the collapse of one of the combatants in the world power struggle, a unique opportunity was created for a radical solution to the nuclear-weapon issue. But instead, the nuclear arsenals are being maintained, albeit at reduced levels. The main reason given for this is that nuclear weapons are needed as a safeguard the potential threat from new nations acquiring these weapons.

Horizontal proliferation is a real danger, but the retention of nuclear weapons as a means of dealing with it is about the worst possible answer. At the heart of horizontal proliferation is the perception that nuclear weapons confer power, prestige and protection. This motivated the earliest proliferators, France and Britain , and is sustained by the fact that the only five permanent members of the Security Council, with right of veto, are the five nuclear weapon states. As long as this nuclear cult exists, as long as the belief is sustained that nuclear weapons bestow status, strength and security, the pressure to join the club will be irresistible.

The main instrument to prevent nuclear proliferation is the Non-Proliferation Treaty. By now 162 states have signed the Treaty, including all five nuclear weapon states. But the NPT is an interim arrangement, step towards nuclear disarmament, as clearly stated in its Preamble. A stable world order must be based on the rule of law, and one cannot imagine an international law that permanently discriminates between nations. If some states are allowed to keep nuclear weapons, because –they claim—they are needed for their security, one cannot deny the acquisition of these weapons to other states.

In the long term, there are only two alternatives: allow the possession of nuclear weapons to all states by eliminating these weapons. There can be no doubt that the former would lead to a highly dangerous, unstable world. The creation of a nuclear-weapon-free world is therefore essential for peace and stability.

A nuclear-weapon-free world is also called for on moral grounds. The whole fabric of civilized society is based on moral values, and if these are violated in one important area, how can they defended in others? Security achieved by the threat of wholesale destruction, possibly genocide, is bound in the long term to erode the ethical basis of civilisation.

Several arguments have been advanced against the idea of a world without nuclear weapon. One is that the genie is out of the bottle and cannot be put back. Nuclear weapons can, of course, not be disinvented, but this does not mean that we have to keep them in perpetuity. It is a hallmark of civilized society that it can control—by national legislation or international treaties—the undesirable products of science and technology.

It is on these grounds that biological weapons have been banned, and a similar ban on chemical weapons has now been agreed to; the Chemical Weapons Convention has been signed by 156 States and comes into force in 1995.

Another argument is that nuclear weapons have kept the peace in Europe since 1945. This is supposition without proof, but has gained credence only by constant repetition. It ignores 125 wars, with over 40 million deaths, in other continents; in Europe too we now have a bloody war. It also ignores the fact that during the past four decades there has been a relentless arms race that has resulted in obscenely huge nuclear arsenals. A more serious objection is that a treaty to eliminate nuclear weapons could be violated by a state concealing a clandestine nuclear cache, or by a later "break-out". Considering the enormous destructive potential of these weapons, such action might give the transgressing state vast power. However, this is not an insurmountable obstacle.

Even with the present state of technology, it is possible to design a system of verification that will greatly reduce the chances of undetected violation. This technological verification can be enhanced by "societal verification", that is by calling on the whole community, to report to an international authority any attempted violation of an international treaty. To be effective, this would require a clause in the treaty, and indeed in national legislation, to make such reporting a citizen's duty.

A recent Pugwash study of such schemes, as well as of methods of enforcing treaties in a nuclear-weapon-free world, concluded that the problems of ensuring the integrity of a treaty to eliminate nuclear weapons are less difficult than is generally believed. The study has shown that more international intervention will be needed, such as control of all fissionable material and enforced legislation such as guaranteed protection of whistle blowing. Measures like these will constitute infringements on national sovereignty, but limitation of sovereignty will have to be accepted in any case. We live in an ever more interdependent world, and the time has come for an extension of the loyalty to one's nation to a new loyalty, a loyalty to mankind.

Mikhail Gorbachev whose adoption of a new way of thinking has transformed the world, was the first contemporary world leader to realize that a nuclear-weapon-free world is an integral part of stable peace. He suggested the year 2000 as a target date, but this has to be understood to mean the time for agreement on a treaty, rather than for the actual destruction of the weapons, that will take many years.

The main task for the remaining year of this century is to convince world leaders, and the general public, of the necessity of a treaty to eliminate nuclear weapons, binding all nations. During this period we should also seek the implementation of intermediate steps, such as a comprehensive test-ban; adoption of the no-first-use policy; tightening of the safe-guard of the International Atomic Energy Agency: and Strengthening the peace-keeping and peace-enforcing powers of the United Nations. An accelerated programme of dismantlement of nuclear warheads, and further deep reductions of nuclear arsenals are of course essential steps.

The very first resolution of the UN General Assembly unanimously called for the elimination of atomic weapons. At long last, the UN is in a position to fulfil the functions for which it has been set up, and the time has come to implement its first resolution; the time has come for a decision to create a nuclear-weapon-free world.

5

Nuclear Arms Race on the Subcontinent

For peace lovers it was strange and bewildering sight; people dancing and cheering in the streets of New Delhi and Islamabad because their government had exploded an atomic bomb, politicians bragging about the nuclear capabilities of their countries, the press going wild over the achievements of their scientific institutions. Achievements? In Europe, people had lived for decades under the threat of a nuclear holocaust, they had danced and cheered when the Cold War ended. Now another threat of nuclear war, this time between two of the most populous, but also the most impoverished nations of the world? How could anybody be happy about the news of the successful nuclear tests? Wasn't it absurd that the masses were cheering when their governments were spending the little money they had on the military and on expensive atomic gadgets instead of combating poverty in their countries?

The political reactions were quick to follow. A day after India had exploded its first bombs under the Rajasthan desert, In a concerted action, the EU and the G7 countries as well as the World Bank suspended all new loans for the country. Pakistan met the same fate after it conducted its own atomic test series a few days later. Development cooperation with the two countries has thus been dealt a severe blow. In the last 50 years, both states have been important recipients of world aid.

The worldwide protests and the suspension of development cooperation with India and Pakistan because of the atomic tests may be interpreted as another example of Western hypocrisy.

People remember that not long ago, the French government was the target of worldwide indignation because of their series of underground nuclear tests in New Caledonia. Stubbornly, the French president at the time rejected all criticism with the argument that the test were necessary for the security of France and that tests would end as soon as enough scientific data had been collected. The Chinese similarly displayed total indifference to world opinion when they followed with their own nuclear test series. Not without justice, the Indian and Pakistani governments point out that they have not signed the Nuclear Non-Proliferation Treaty and the Test Ban Treaty and are, therefore, not bound by any international commitment to observe nuclear abstention. India and Pakistan can also rightly point out that the nuclear powers have not fulfilled their own commitments for nuclear disarmament, which are part of the Non-Proliferation Treaty. It is really difficult to explain why the five members of the exclusive nuclear club should have the sole right to nuclear arms including the right to develop ever more sophisticated and deadly nuclear weapons. Nobody can be surprised that big countries like India are demanding equality with the nuclear 'haves', and that a hostile neighbour like Pakistan which sees itself as the rival of India on the subcontinent is trying to stay in the race by building its own bomb.

This is the political side of the coin. The quest of governments to 'keep up with the Jonesses'. Their claim not to be a second class power. But there is also the other side of the coin: Is nuclear equality really in the interest of the people (even if they are dancing in the street when the bomb goes off)? What do they gain from it—for their daily lives, their health, their education, the future of their children? Do their lives not become more insecure, threatened by nuclear war or nuclear accidents? The people in India and Pakistan must answer these questions for themselves. At the moment, gripped by chauvinistic excitement, a majority seems to believe that national aggrandisement is more important for them than better schools, hospitals, houses or roads.

If Indian and Pakistani scientist are able to build nuclear bombs, are they not also able to solve other technical problems in their countries? If their governments have the money for nuclear armaments, should they not be expected to also pay for infrastructure projects, for the provision of fresh water and electricity, for the running of health and education services? How can foreign donors be asked to deal with poverty reduction in the slums or rural areas of their countries, while the military claim a rising share of the national wealth for their ambitious armament programmes?

Certainly, India and Pakistan are sovereign nations and as such they have they same right as the established nuclear powers to spend their money on the atomic bomb. But if they choose that option, they should not ask other countries to help them in their social and economic development. Development cooperation already suffers from a waning acceptance among the population of donor countries. The Indian and Pakistani bombs have given "development fatigue" another push. People should know this when they rejoice over the arrival of their countries in the nuclear club.

6

A Crucial Encounter

Genetic tests and treatments must not be allowed to create new forms of discrimination between those who, for whatever reason, can or want to take advantage of them, and those who cannot, mostly for lack of money. If a scientific discovery can form the basis of a technology, then it is highly probable that the technology will eventually be applied. Today this lesson of history is causing anxiety among politicians, scientists and public opinion concerned about the current far-reaching developments in biotechnologies.

It is now possible to penetrate to the very essence of living things as a result of spectacular scientific advances that are gradually revealing the innermost mechanisms of life. The technologies based on this field of knowledge offer humanity for the first time astonishing power to revolutionize the process of creating and developing human beings, and ultimately, the human species. Technically speaking these breakthroughs could lead to the revival, in even more effective guises, of eugenic practices we hoped had been buried forever, Fortunately, this nightmare scenario seems highly unlikely.

But history also shows that new technologies are rarely applied without a framework of rules and procedures designed to ensure that they are beneficially used. Human progress has always been driven by the winds of freedom, including freedom of enquiry and initiative, but human beings have always tried to head in the right direction and to respect certain limits. The biologists have done their work; they have sown the seeds of vast possibilities. Now it is up to society to make sure that only the benefits are harvested. The biotechnology revolution beckons humanity to a crucial encounter between science and ethics.

Where human reproduction is concerned, as with technology in general, we must be guided by respect for three basic and interdependent principles dignity, freedom and solidarity.

For human dignity to be respected, each person must be regarded as unique. This position has far-reaching consequences for human procreation first of all, it rules out cloning as a means of reproduction because this technique, which is almost upon us, involves genetically "duplicating" an existing person. More generally, predetermining the basic characteristics of a future person, notably trying to enhance their future physical or mental capacities, violates the very essence of human individuality. This kind of engineering would end up by depriving individuals of that which is theirs alone—the mysterious processes whereby their unique genetic heritage emerges and interacts in its own unique way with their environment.

Advances in prenatal scanning and testing techniques may confront parents with grave new decisions. The danger is that various kinds of pressures or even regulations will develop which only allow "genetically correct" people to be born. This would be totally unacceptable. No authority—be it political, social or economic should be able to enact such a "genetic order", still less impose it.

So increasing emphasis must be laid on solidarity. Genetic tests and treatments must not be allowed to create new forms of discrimination between those who, for whatever reason, can or want to take advantage of them, and those who cannot mostly for lack of money.

The risk of uncontrolled, unmonitored genetic engineering increasingly looms over us. But we are starting to see the emergence of a new "responsible" form of genetic engineering in which the power of science is subjected to the power of ethics an ethics that benefits everyone, not just a few, and looks towards future generations, not just short-term interests.

7

A Universal Responsibility

Each of us is responsible for replacing the logic of force with the logic of reason and respect for the view of others. On the threshold of a new millennium, the issue of responsibility is taking on a new dimension. Humankind is still beset by war·and violence. It also faces new global challenges. The impact of human activity on our planet is so great that for the first time in recorded history, we may be approaching the point of no return The widening asymmetry within and between countries, environmental destruction and flourishing arms sales raise doubts about many of civilisation's values and standards. How can we handle new global threats? History shows that no situation is hopeless if risks are identified early enough.

The conflicts that have arisen since the end of the Cold War have erupted not as a consequence of new freedoms but in reaction to earlier oppression or repression. Suspicion intolerance and hatred built up over decades, even centuries. But alongside the armed strife of recent years, humankind has begun to demonstrate a new skill in resolving conflicts. Mozambique, El Salvador, the Philippines, the changes in South Africa that would have been inconceivable just a few years ago, the efforts for peace in the Middle East and finally, the beginnings of a settlement in Northern Ireland are all examples proving that conflict is not inevitable. They demonstrate that breakthroughs to peace can be made by dialogue, mediation, negotiation and imagination not force.

That is why the transition from a culture of war to a culture of peace is the foremost challenge as the twentieth century draws to a close. To succeed we—all of us, day-in and day-out must not only do away with approaches based on force and imposition, but profoundly change cultural attitudes and daily behaviour.

We must use imagination and resolution to go to the roots of world problems and nip conflicts in the bud or, better still, prevent them. Learning to live together means daring to share, daring to do things differently, and daring to dream of a better, safer, more just and human world. It also means having the resolve and courage to transform our dreams into reality.

Here I wish to underline the pivotal role of education in promoting a culture of peace. By education, I mean not only formal instruction in schools but also informal training within a whole range of cultural institutions, including in the very first place the family and the media.

Who will be responsible for changing the culture of war into a culture of peace? Governments, parliaments, intergovernmental organisations, we might reply. The answer is correct, but it is not the only one. The transformation cannot be achieved without the active involvement of those with financial resources and influence. This answer would also be true, though only partly so. For in the final analysis, replacing the logic of force and confrontation with the logic of reason and respect for the views of others is a responsibility that belongs to all nations and all citizens, to each of us no matter how great or small the scope of our individual responsibility. The challenge of promoting a culture of peace is so broad and far-reaching that it can only be accomplished if it becomes a priority for the entire United Nations system. The dream of achieving a world without strife and violence is urgent.

The Mahatma Gandhi said, "In the midst of darkness, light prevails." It is that light which is spread by the democratic values enshrined in our constitution: justice, freedom, equality and solidarity.

8

Science to What Purpose?

Are the welfare and interest of the public being served by the priorities of researchers, the thrust of their, work, the ways in which they are organized, the funding they receive, and the circulation of their findings?

Science reigns triumphant. Never has it been so powerful and influential. It has conquered diseases, which have decimated whole populations. It has abolished exhausting physical labour and wearisome repetitive tasks. It has vanquished distance and pushed back the frontiers of our knowledge of the infinitely large and the infinitely small, in both the inanimate and the living world.

In short, it has acquired the ability to shape our lives, to change life itself. But it has also increased its capacity to destroy life. The strength of an army can rest on the number and determination of its combatants but it is also, and chiefly, based on the technological sophistication of their weaponry. The bombing of Iraq, and now of Serbia, are the latest examples.

Yet science is wavering. For the first time since the Enlightenment, the way science can be used is being challenged. The link between scientific progress and social progress is weakening and signs of obscurantism are appearing. Hiroshima sounded the alarm. Then the crisis of the environment, triggered by the dominant mode of development, turned questioning of science into a worldwide issue. This form of development is inseparable from a frantic and indiscriminate quest for technological innovations. Finally, advances in biotechnology, which harbour many grave dangers for human dignity, are often too closely bound up with the selfish interests of their promoters.

No one blames science for not knowing everything. No one criticises it because it has not yet found a vaccine against AIDS or reached a conclusion about the theory of the Big Bang. It has never been claimed of science, as it has of history, that it has come to an end. It must keep on tirelessly probing the enduring mystries of life.

But science can no longer avoid—and nor can we—the basic question of what and who it is for. In other words, are the welfare and interests of the public being served by the priorities of researchers, the thrust of their work, the ways in which they are organised, the funding they receive, and the circulation of their findings? Or are scientists looking mainly in the direction of high-spending consumers at the expense of long-term basic research? Because of the growing "privatisation" of research, are we not tending to overlook essential and universal human needs, which cannot immediately be met?

Those who are excluded from this new "scientific power" must make their voices heard. For example, the inhabitants of the 600,000 villages which have no electricity or the world's two billion people without access to drinking water have the right to ask science to find solutions adapted to their very meager resources. Humanity also has the right to ask science to give priority to research into processes of global disruption and ways of coping with them. What's more, all citizens have the right to ask science to further our understanding of the mechanisms of inequality and exclusion which are gradually undermining peace and democracy.

One major purpose of this paper will be to see that the benefits of science go primarily to all those who have hitherto been unreached. Their conditions will only improve if they have access to the mighty power of science.

9

The Fabric of Peace

In a world pervaded with violence, the struggle for peace must begin in everyday life. From time immemorial, peace—in the sense of peace between nations and peace within societies—has been exclusively based on the interplay between justice and force, a relationship which is at once conflictual and consensual.

History shows that peace has been and continues to be primarily an affair of state and of states, based on the use of force—in other words, ultimately, on recourse to war. This force is legitimized by highly diverse and in some cases contradictory concepts of justice.

In our time, however, the nature of war has changed. Most often it is no longer waged between states but within their borders. Conflicts within states have become so widespread that warfare has never been so rife—and in many cases so unnoticed –as it is today. Furthermore, violence exists in all our societies in one form or another, even if it does not necessarily erupt into armed confrontation. It takes many forms, and may even been considered the norm. Its most glaring symptom is the growth of inequalities and the social exclusion to which this gives rise. Societies torn to the point of disintegration by civil war or violence are societies whose regulatory mechanisms—the bodies which exist to settle conflicts—are wrapped or paralyzed.

Some might say that this gloomy analysis heralds a future in which war and violence will inevitably prevail. The fact is, however, that these three developments—the changing nature of war, the proliferation of different forms of violence and the weakening of mediatory mechanisms, a process accelerated by globalisation and the information revolution—create a new arena in which a culture of peace can emerge. And the prime mover in this culture of peace will no longer be the state but the individual, in other words each and every one of us.

For surely the road to peace must start within ourselves—in valves, behaviour and attitudes which can foster a sense of community that is today threatened. Where else can the foundations of peace be built but in our daily lives, through willingness to listen and talk things with others on equal terms within the framework of a caring society?

The only obstacles to this enterprise are those we create ourselves, because of ignorance or fanaticism or because of the selfishness that today we are all too often asked to regard as the hallmark of human identity. This approach requires more than good intentions or the occasional act of generosity. The capacity to talk and listen to others and be receptive to their needs can pave the way to peace through an acceptance of a shared responsibility towards other people as well as towards ourselves. The mainspring of the culture of peace is making common cause with others in peace-building projects in everyday life, in whatever area of society we may be involved.

A Participatory Process

This message is not new. The culture of peace is a fabric that has been woven for generations in all societies, though its practices are not necessarily dominated by that specific title. In some places it may be known as tolerance, non-violence or justice. In others, as harmony, solidarity or conviviality. All over the world it has its defenders, some working in obscurity, others in the spotlight of public life. But its scope would be much smaller today had it not been given expression in the disinterested acts of thousands of anonymous men and women capable of listening to others, talking to them and acting with them and on their behalf.

The concept of "a culture of peace" has clearly not appeared from nowhere. But to have a single expression to describe a multigrade of ethical and practical initiatives may help to highlight their common purpose, make them more widely known and bring them together. It may sharpen the impact and focus of movements that are active in a vast range of fields and settings.

The importance of the culture of peace has now been recognized by the world community. The General Assembly of the United Nations unanimously proclaimed the year 2000 as the International Year for the Culture of Peace. In taking this step, UN Member States accept their own limitations and their urgent need for concept of peace that will be a participatory process to which all members of society can contribute no matter how humble their circumstances.

The culture of peace is intended to be a rallying point that transcends the treaties and agreements that have so often been given short shrift by history. It will become a real and living culture if we take its tenets to heart and shape a common future in our words and deeds.

Peace based exclusively upon the political and economic arrangements of governments would not be a peace which could secure the unanimous, lasting and sincere support of the peoples of the world, and that the peace must therefore be founded, if it is not to fail, upon the intellectual and moral solidarity of mankind.

10

Trading Towards Peace

The reason why trade has such a vital part to play in building peace is because it means lowering barriers—not only to goods and services but among nations and peoples. The elimination of barriers creates interdependence and interdependence creates solidarity. The history of the last fifty years has shown us all the undeniable benefits of lowering trade barriers and opening economies.

Clearly every region has its own characteristics, and it would be wrong to imagine that the same blueprint can apply everywhere and in the same way. Any region which was for thousands of years at the crossroads of world trade should regain its place in the centre, because doing so will help build peace as well as prosperity. This is why the numerous applications for accession to the WTO from various countries are so significant. The first is through regionalism. There are several efforts at regional trade and economic initiatives among countries, and that such initiatives will be encouraged to reduce positive results. Regional initiatives are important because they can help countries at a comparable level of development to move relatively quickly in opening their economies and in deepening their interdependence.

However, the rapid advance of global economic integration means that while regional initiatives remain important, they are not sufficient by themselves to address successfully the new perspectives of the international economy. That is why there is a need for second track, which is the rule-based multilateral system. And that is why the multilateral system is of fundamental importance to the economic prosperity of any region.

As the first major international institution to be created in the post-Cold War era, the WTO offers a promise of the kind of global economic architecture which need in the coming decades. Its culture is firmly rooted in the tradition of consensus-building and cooperation among sovereign countries. And the WTO embodies rights and obligations negotiated by consensus, approved and ratified by each government and each Parliament, and they are enforceable, not through the crude exercise of economic power, but through the rule of law. The alternative would be a power-based system—who would want to chose this option?

But most importantly, the WTO is an organisation which brings all countries—from all corners of the world and from all levels of development-together as equals. There is no weighted voting, no exclusive clubs, no inner and outer circles. Developing countries representing 80% of constituency sit as equals with industrialised countries to write the rules of a shared trading system.

This new unity of developing and developed countries inside a single system will be credited as the greatest achievement of the multilateral system. But this unity is still fragile: We cannot allow it to be broken: This is why, in preparing the agenda of the first Ministerial meeting in Singapore, have recognised the particularly difficult task facing developing countries in implementing the Uruguay Round commitments. They have also acknowledged the challenges they face in contemplating the necessary work programme.

The integration of developing countries as equal partners in the multilateral system is one of the most important challenges in shaping the economic order of the 21st century. This is a shared responsibility of developed and developing countries alike. There is no rational alternative to this objective. The evolution of the global economy makes that clear.

Now there is a need to work together as equal partners to ensure the full integration, and all other developing and transition economies, into the global economy and the rule-based multilateral trading system. In conjunction which this there is a need to encourage, notably the growth of regional economic cooperation. The alternative is a vicious circle where economic isolation feeds greater political instability which in turn leads to greater economic isolation. The road to a lasting peace in the world begins, not ends, with economic integration and interdependence. Taking this message to heart will help build a future where it is goods, services, and investment that cross borders—not missiles and soldiers.

11

Peace and Poverty

Peace should not be understood in military terms, like absence of armed conflicts. Peace should be understood in a human way in a broad social, political and economic way. Peace should mean social justice between nations and within nations. It should mean establishment of human rights for all people.

In the new context the concept of "peace" would be the existence of a political and economic environment where each individual human being is truly free; free from the control of any powerful person or any powerful nation, free from poverty, hunger and indignities, each individual human being free to explore the limits of one's own potential.

Today peace is threatened, more than anything else, by poverty, unjust social and economic order, absence of democracy and environmental degradation.

The cold war cloud has gone. You can feel the breath of fresh air around the world. Now there is no visible competitor left for capitalism. It is quite risky to live with a philosophy, which has no challenger. To be safe, we must go to the essence of the philosophy of capitalism rather than be satisfied with the practices, which emerged over years through patchworks of expediency.

Contrary to common belief, it is not the "free enterprise" which is the essence of capitalism. It is the freedom of individual thought and freedom of individual action, which is the essence of capitalism. It is these freedoms, which support free enterprise, free trade, free circulation of capital, and free circulation of people.

We must work out a new system, appropriate for the new world, from the basics of capitalism, not from the practices of capitalism. Many of these practices take away freedom, rather than guarantee it. Traps must go. People cannot remain trapped in places where they cannot live because of ecological, political, or economic reasons. This planet belongs to all people. If some people are trapped somewhere, we must all come forward to remove the causes of their discomfort. At the same time we must leave our shores open for anybody who decides to join us, or anybody who decides to part our company.

Poverty denies a person control over his destiny. Poverty means not being able to tell what tomorrow would be like. If we examine the situation carefully we'll see that the poverty is neither created by the poor, nor sustained by the poor. It is the system of policies and institutions that we have built around us that creates and sustains poverty. Poverty is the denial of human rights. Over one billion people live below the absolute poverty line right now on this planet, are denied of almost all human rights. There is no way one can defend the existence of poverty anywhere. Poverty is a disgrace for the entire man—kind. Because we allow another human being to die of hunger, or malnutrition, or common curable diseases, or exposure to climate, we are reduced to less human being. If a particular world system is responsible for creating this massive poverty we must act to replace it.

Resource-wise or technology-wise, there is no reason why poverty should exist and continue to deepen and widen. If we make up our minds to wipe out poverty from the surface of the earth, the worst aspect of poverty can be removed within the next couple of decades.

We can build a poverty-free world at a fraction of the cost of what we spend on war preparations. Nations become very generous when it comes to making their war-machine heftier in the name of ensuring "peace". Can we persuade ourselves to allocate a part of our time, money and intellect to achieve peace by making the people at the bottom the winners, rather than nations winning wars? "Peace" achieved by winning wars is earned by destroying people. The real peace can be achieved by building people, by reinforcing people, by helping people to reach their potential. Removing poverty is the process of building people.

Each human being is a wonderful creation of the Creator. Each human being is born with great potentials. Poverty denies any opportunity for a person to achieve any of his/her potential. We have built a world system, which is in the habit of pushing people down not building them up. It creates barriers around individuals, rather than remove them.

The most effective step that we must take to remove poverty is to create a system, which creates enabling conditions for people and removes the existing barriers. The institutional barriers were skilfully crafted over the centuries to benefit a handful of people.

Resource—poor nations with high incidence of poverty waste away enormous human capability each day by denying poor people the use of their energy and ingenuity. If they could have been made economically active, not only they could have contributed in the national production, they would have helped expand the domestic market for the products produced. The poor can be transformed into the engine of growth if we only allow them to unleash their capacity.

We cannot be at peace with ourselves if we know there is a human being who lives a life worse than an animal. A human being is supposed to live differently than an animal. He/she is supposed to live a life with human dignity. Human dignity is what distinguishes a human being from an animal. When we cannot ensure this dignity for others, our own dignity becomes an empty pretence.

There must be a thousand and one ways to remove poverty from the earth. We may or may not know some of those ways already. Obviously there are many more ways yet to be designed, each more effectively than others. When we shall find them, how many of them we shall find, how quickly we find them; will depend on how eager we are to find them. But to say that poverty cannot be overcome, directly and quickly, is to underestimate the capacity of human mind.

Poverty is homogeneous only when considered from the point of view of income or consumption: the uniformity of the poor as a category exists only on the level of the fact that they have little to consume. When considered from the point of view of production, i.e., the circumstances in which the poor must operate to gain their income, the conditions of poverty are extraordinary diverse. A concrete grasp of these diverse circumstances is the first step in developing relevant instruments to address not only the problems of the poor, but also the challenge of taking advantage of the opportunities available to them.

The conventional means of measuring economic progress, such as Gross National Product per capita, tell us little about the real nature of poverty. In recent years this sort of yardstick has been supplemented by measurements of food security, income distribution, and social development (encompassing health and education). These offer the possibility of composite indices, allowing the development of more rounded characterisations and comparisons of poverty at the national level. However, these principally refer to the symptoms of poverty, not to the relational factors generating it. Poverty is not a state of being; it is the effect of dynamic processes. While it is important to know where poverty is greatest, it is critical to know why it exists. This inquiry necessarily leads away from the nature of the poor as individuals to the nature of their social and physical environment. Poverty is not only a personal phenomenon, it is a social status. As such, while its effects can be measured on the level of the individual, its causes must be sought elsewhere. From the point of view of poverty alleviation the process of becoming is just as important as the state of being.

At the heart of poverty is the inadequate access of the poor to productive resources. Low incomes tends to reflect inadequate means of production, not incompetent producers. However, poverty in India is not simply a reflection of private resources. A broad range of "external" factors impinge on incomes, among them the following:

National Policies

One of the ironies of Indian development is that while no government wants poverty, many policies contribute to it –what is given in anti-poverty programmes is drained away by other policies. The poor do not always come out ahead in the balance—they are often net "donors" to the rest of society. Frequent reference is made to unsustainable forms of development—to urban over-expansion, industrialisation based on subsidies, and to public sector engorgement. What is less frequently realised is that the bill for these phenomena is often presented to the rural poor. Taxation of exports to sustain sectors with little export potential of their own and subsidized food imports to supply the urban population are policies that are often paid for by the rural poor. In many areas of India, exports are agricultural goods produced by small farmers. Here export taxes contribute to rural poverty. The same is true of "cheap" food imports, which depress the prices paid to small farmers for their food crops.

"Structural imbalance" is not only a recipe for increasing external indebtedness; it is also a recipe for increasing the poverty of the rural population. The political weakness of the poor in most areas is not only the basis for inadequate poverty alleviation programmes and policies—it is the basis for an actual transfer of their income to more socially influential groups. While it is often correctly asserted that the poor are the first to suffer from adjustments involving public social expenditure cuts, it is often the case that they also have the most to gain from the elimination of policy-based economic distortions that reflect social power rather than productive efficiency and potential.

Demographic Factors

Accelerated population growth is a long-term contributor to poverty. In India the incomes of the poor have declined, mortality rates are also falling, pushing the numbers up. In the meantime, land is becoming scarcer, plots more fragmented and the soil and pasture increasingly degraded. This phenomenon is not without its policy dimensions. As long as the poor remain undercapitalized, and essential determinant of household income is the amount of labour available to its household economic strategies favour large families. While population policy has a role to play, possibly more critical is a change in the economic environment. Access to capital and more secure income changes perceptions of the need for labour. In the medium—and long-term, population dynamics are driven by the underlying productive systems. As long as the production systems of the poor remain underdeveloped, population growth remains high, restricting even the future possibility of development.

Natural Resource Management and the Environment

If poverty is both cause and effect of rapid population expansion, so poverty is both cause and effect of many dimensions of degradation of the environment. Many of the rural poor, but by no means all, live in areas of extreme environmental fragility, a circumstance often prompted by high level of control by the better-off over more stable and productive resource areas. Here the poor are extraordinarily exposed to the dangers of erosion, whittling away at an already meager productive base. The threat is not entirely due to nature. Rather, poverty accelerates erosion. Without capital, the poor are frequently unable to invest in even traditional methods of soil and water conservation. And without sufficient land they are forced to shorten fallow periods, putting further strain on the resource base. As in the case of population growth, the result is strain not only on the poor, but on the entire Indian economy. Given the extremely limited economic alternatives, the solution to this problem is not to forbid the use of environmentally fragile resources to the poor; it is to change the conditions under which their use takes place. Access to conservation technology is important; but more so are security of land tenure and resources to invest.

Combating poverty means not only increasing the production of the poor, but also preserving and enhancing the long-term value of the resource they control. What this very often means, in practice is assisting the poor in reestablishing a stable relationship with fragile resource. Prevailing processes in many areas involve the gradual—and sometimes not so gradual—depletion of natural resources, to the detriment of all. Part of he answer to this is conservation. Part of the answer is also to provide viable economic alternatives to the poor, reducing their dependence on erosion-prone crop and livestock practices.

Exploitative Intermediates

The poor are not unaware of the pressure upon them, and also of means of overcoming them. Their ability to respond, however, is severely impaired by social powerlessness. The poor are surrounded by a dense network of public and private factors reducing their freedom of action, and actually draining what few resources they do have. Members of the network include traders and moneylenders capitalizing upon the economic weakness of the poor, and engaging them in unequal exchanges. They also include public agencies either indifferent to the requirements of the socially uninfluential, or actively engaged in extracting "surplus" for use by other groups. Not to be excluded from this are organisations which are ostensibly "for" the poor, but which, in fact, serve as systems of containment and control.

12

Free Trade as Peacemaker:
The Benefits of an Open World Trading System

Globalisation by free trade according to the principles of the World Trade Organisation (WTO) offers the only realistic opportunity to integrate the world peacefully and in time to prevent a major disaster. The primacy of the economy over politics is the most important vehicle for a successful world domestic policy.

Since Adam Smith, traditional economic theory has on principle been well-disposed towards free trade. Free trade enables better use of the world's economic resources than does national protectionism. Countries can concentrate on their respective strength and draw from their trade partners the goods they need, but do not produce. But there have always been objections against free trade.

The international trade system has always been encumbered by disparate accusations of unfair methods, such as dumping, as and is widespread. If one were to believe all the charges of dumping that are made, then international trade would have been completely destroyed long ago. Great restraint should be exercised with respect to allegations of dumping if one is interested in maintaining an interweaving of international economic activities.

The Free Trade Opposition Cloaks in Dumping Charges

The modern form of the struggle against free trade cloaks itself in the accusation of ecological dumping or social dumping.

With this difficult subject matter, one should not make sweeping generalisations. These things also are not gone into in detail in what follows.

Environmental protection is an asset that every economy produces at the cost of other assets. The people's preferences for the asset of environmental protection probably varies from country to country. It is also completely legitimate and does not at all distort trade if the environmental provisions—in line with the different national preference—vary from country to country.

In the rich western European economic region, one should guard against a new form of cultural imperialism. It is not for this part of the world to impose its preferences for environmental assets on other countries, especially Third World countries. Free world trade brings not only economic advantages. Even more important is its contribution to lasting world peace.

In view of world population growth, every standstill in the movement towards a peaceful world society must be seen as a step backwards. We are compelled to run a race between the growing problems and the development of stable institutions to overcome them peacefully at global level. Economic history since the end of the Second World War shows clearly that free trade under the old GATT was of decisive importance for the prosperity of the industrialised nations.

The principle of help for self-help has nowhere been applied so consistently as on the free world market. In reverse, the example of countries that cut themselves off from the world market show the disastrous consequence of the rigidity of a society which shuns the pressure of international competition.

Revolutionary Success of Open-Market Policies

The West's policy of open markets pursued since 1948 and reinforced since 1989 has led to dynamism, which, in the true meaning of the word, is revolutionary. More than half the world population now lives in countries with annual GDP growth rates of more than 5 per cent. Europe is not among that group, which may be why it also stands somewhat apart in its mentality.

Certainly, there also can be undesirable trends in free trade. There is no ideal systems; one must choose between imperfect potentialities. However, no realistically better substitute for the free trade system is in sight, not even with respect to the goals of pacified world; an ecological sound world economy and a balance of global dimensions between the poor and the rich. An ideal government of philosopher kings armed with absolute power certainly could do some thing better than does free trade—but such a government remains fictitious. There are tangible and narrow limits to what the political system, whether democratic or not, can effect in a positive sense. This is how the structural conservatism of democratic and other political systems impedes the timely assertion of reforms necessary to achieve a world peace society.

The GATT was turned into the World Trade Organisation (WTO) a few years ago. Besides extending the free trade principle to services and additional agricultural sectors, the new agreement foresees above all the full inclusion of the Third World in the system. The agreement commits the industrialised nations to open their markets to developing and threshold countries.

Other important points are the strengthening and tightening of the dispute mediation process. Based on a system of relatively independent ad hoc panels, it permits complaints against WTO member countries for violations of agreement. Thus, what is arising here is an effective global jurisdiction within the meaning of a peaceful world domestic policy.

Exclusion as Penalty

The decisive sanction mechanism of the WTO—which is not a specialist organisation of the United Nations—is the threat of exclusion. Exclusion would deny the penalized country free access to the markets of WTO members on the basis of most favoured nation status. This is a threat that requires no armed force, but is very effective. No country can still afford to do without the beneficial effects on prosperity that participation in international trade brings.

Thus, with the threat of denial of access to world markets for violating WTO rules, and the guarantee of a more or less fair competition for a country's own products for abiding by them, a non-military sanctions system has come into being. That is substantial progress on the path to a pacified world.

Certainly, this sanction system's sphere of influence is limited for the time being. Essentially, it will be used to assert the game rules of free trade. It offers no legal grounds for pressing other goals, such as on human rights, Attempting to expand it in this direction would for the foreseeable future put the entire system at risk.

In the current debate on globalisation, the question arises of whether the world economic institutions should not be converted in this manner, that politic regains its autonomy, and that the primacy of politics can be restored. The critics of globalisation point to the constraints to adjust which the world economy exercises on national or continental politics. However well this demand for the primacy of politics may be justified in philosophical terms, it virtually comes down to a demand for the ascendancy of the conservative principle.

Danger of a Slowed-Down World Integration

The demand for the primacy of politics is gaining strength from the desire to avoid the pressure to adjust which the dynamics of world events are exerting. It is today a conservative, and in fact a reactionary, longing for the (Utopian)return of the functioning European welfare state of two or three decades ago. If it were asserted, it would mean practically slowing down world integration. It would run dead against the goal of a world policy based on a desire for peace.

The present primacy of the economy over politics in terms of the free movement of goods, services and capital—is basically nothing more than the priority of the global principle over the provincial, the national principle. As such, it gives the principle of change pre-eminence over the principle of maintaining the status quo. What gives the primacy of the economy its legitimacy? Probably not the thought that world peace can better be secured by this means. Its legitimation lies in the very indirect economic success that the free trade system delivers. For the reflective observer, the question remains of whether this legitimation is sufficient.

To answer this question, however, and particularly if one pleads for maintaining the ascendancy of the economy, it appears appropriate to outline the consequences that can be expected from further integration of the world economy. As can be seen today in East and Southeast Asia, the growth dynamics of the world economy will lead to a marked rise in the living standards of a large part of the Third World.

Do not Exclude Poor Countries From the Competition

The global consequences of Asia's growth should not have been seen only negatively. While it also may mean, for example, a great burden on the global climate, it leads at the same time to an acceleration of the process of falling birth rates and thus to an earlier stabilisation of the world population. Prosperity for the Third World is so far the only realistic answer to the urgent problem of population growth. And free competition on the world market is the only reliable means of achieving this prosperity in the course of some decades.

Despite ecological sacrifice in the medium term, continuation of Third World growth is the only way to solve the long-term ecological problems. One also should not forget that only those who can eat their fill and have a roof over their heads are prepared to reflect on ecology and discuss it.

As for the rest, the balance between rich and poor is more acceptable when the poor become richer than when the rich become poorer. That applies also at the international level. The market and access to it are peaceful sanctions of the world economic system on the basis of free trade. Those who are hungry and have nothing more to lose are more of a danger to world peace than those who have eaten their fill. The ruse of covering up domestic problems by cross border military aggression will become less attractive to the degree that a country's own economy is integrated in the global economic system. The more countries are economically dependent on each other, the more unlikely it is that they will wage war on each other.

Globalisation by free trade according to the principles of the WTO offers the only realistic opportunity to integrate the world peacefully and in time to prevent a major disaster. The primacy of the economy is the most important vehicle for a successful world domestic policy.

13

The End of the Old Order:

No Guarantee for Peace and Prosperity

In the last few years, the world has witnessed the collapse of communism and the end of many authoritarian regimes around the globe. In Eastern Europe and the former Soviet Union, multiparty systems with free elections have been introduced. In Afghanistan and Cambodia, Ethiopia and Angola, Nicaragua and Peru leftist government of various shades have given way to more pluralistic forms of government or are in the process of doing so. In south Africa, the white minority has accepted the black majority rules. All over the world, the old order which was established after World War II is breaking down. Freedom, Democracy, self-determination and economic prosperity are the slogans of the revolutions, which have swept the repressive regimes away.

But after the initial euphoria over the unexpected successes of the democracy movements, the world is now waking up to the sobering recognition that freedom does not automatically lead to peace and economic progress. In fact, in many countries that have managed to throw off the communist yoke ethnic rivalries have been sharpened by the call for self-determination. Age-old historical animosities between various nationalities and religious groups have come to the fore and are threatening to undo whatever gains have been achieved by introducing a pluralistic system of government.

The former Yugoslavia is a case in point. The state was an artificial creation born after the First World War and the demise of the Austrian-Hungarian and Ottoman Empires. Similar to many of the artificial states in post-colonial Africa, Yugoslavia consisted of a diverse mixture of ethnic and religious nationalities held together primarily by the charisma of former president Marshal Tito and the socialist ideology he imposed on the country. The wind of democratic change which blew across the European continent in the wake of Gorbachev's perestroyka also meant for Yugoslavia and hence it is divided between different ethnic or religious groupings.

What has happened in Yugoslavia is also taking place, although to a lesser extent, in Georgia, Armenia, Azerbeidjan, and Moldova where different nationalities have taken arms against each other to fight for their right of self-determination. In Ethiopia, the right to secession has been conceded to the Eritreans by the new democratic government, but already the Oromos and other ethnic groups are threatening to break away from the common state. In Afghanistan, the defeat of the communist regime has not brought peace to the country but the danger of a prolonged bloody civil war between the victorious guerrilla factions. Even in South Africa the end of apartheid marred by intensified inter-ethnic fighting between different African parties and groups.

The conclusion to be drawn from this review of recent developments around the world is certainly not that the call for freedom and democracy will inevitably lead to chaos. However, these developments must warn us that the wonderful concepts of Western philosophical thought will only work in practice if they are accompanied by the necessary spirit of compromise. Democracy functions well in societies, which are very homogenous, like many of the Western European ones. Even there the system may fail when faced with deep-rooted antagonisms as the case of Northern Ireland shows. But in countries which are divided by nationality, race, or religion freedom, democracy, and the right to self-determination may lead to even sharper conflict unless there is a genuine give-and-take between groups which grants every individual and every group the right to develop according to their own ideas. The older order is breaking down. The communist regimes have collapsed, most of the military dictatorships are on their way out. But in many parts of the world it is not yet clear what will come in their place. One thing is certain: there is no easy way to peace and economic prosperity. The new systems that emerge will have to guarantee the rule of law and the protection human rights. And they must try to instill a spirit of common values, which bridges the cleavages, which separate ethnic and religious groups. The oppressive regimes which held together their populations by force must be replaced by government which have learned the art of compromise—which is the essence of pluralism and democracy.

14

One Battle After Another

Women fought for their rights throughout the twentieth century. In the past several decades, their struggles has truly become global, but all is far from won. We often hear that this will be century of women, in light of the tremendous strides that have been made in the past thirty years or so. Although it is far too soon to confirm this prediction, it can safely be asserted that the twentieth century was marked by their struggle to leave the home, where they were confined by the ancestral division of roles along gender lines. Around the world, women have campaigned to win the rights they have been denied and to build, side-by-side with men, the future of the planet.

True, such struggles had already been waged in the past, although they were deliberately shunned in official historical accounts. But the brief revolts of this special "minority", which accounts for over half of humanity, did not change the place of women in their societies. They may have ruled the roost, sometimes enjoying undeniable respect, but nevertheless they were still born to serve men and bring their husbands' descendants into the world.

Education: Their First Struggle

Yet, at the start of the twentieth century, the traditional distribution of roles, seemingly legitimised by every religion and frozen in a "natural" order, began to crumble under the two-pronged assault of modernisation and women's struggle for their collective emancipation. They waged many battles to gradually obtain, despite set-backs, a change in their status—which is still far from achieved.

The first struggle of the twentieth century was for education. In 1861, a young woman graduated in Finance with a baccalaureate, a high school leaving examination, for the first time. In 1900, the first female university was founded in Japan. The same year, girls won the right to secondary education in Egypt and the first girl's school in Tunisia. Young women who could made the most of these new educational opportunities, not only to become better household managers and good educators for their children, as the main discourse of the period suggests, but also to do something unprecedented: to enter the forbidden spheres of public life, to exercise citizenship and to participate in politics. Throughout the twentieth century, women waged a battle on two fronts: by fighting for their own rights and taking part in the major social political emancipation movements.

The earliest feminist movements, which first appeared in the West in the late nineteenth century, focused on workplace and civil rights issues. Industry needed women's labour, which was underpaid in comparison with that of their male counterparts. 'Equal pay for equal work!' demanded American and European women, who began setting up their own trade unions and organizing strikes. They made unquestionable strides, but after more than one century of struggle, most women around the world still earn less pay for equal work.

The Right to Vote

The second objective of the twentieth century's pioneers was participation in public life, which hinged first and foremost on having the right to vote. The struggle was long and sometimes violent, as shown by the British "suffragettes" who demonstrated in the streets or Chinese women who made their demands heard by invading their country's new parliament in 1912. Everywhere, the fierce resistance of the political world progressively yieded to determined women's movements.

Control Over their Own Bodies

For a while, women's rights movements took a back seat to the Second World War and liberation struggles in the European colonies., The fight against fascism and, after 1945, colonialism, mobilised all their energy. Women distinguished themselves in these struggles, but that did not sffice to establish their right as a gender. However, the world continued to change. With independence, many women in the South won access to schooling, salaried employment and, in a few exceptional cases, the closed world of politics. In Western countries, the post-war period saw them enter the work force on a massive scale. The gap between social reality and the discriminatory laws defended by exclusively male power structures grew wider.

In the West, the second generation of feminists emerged in the wake of the libertarian movements of 1968. Picking up where their elders left off, they broadened the scope of their demands, for late-twentieth century feminists no longer aspired to the right to be "just like men" Challenging the claim of the "white male" to represent university, their goal was to achieve equality while remaining distinct as women. The women's liberation movement that first emerged in the American middle-class claimed the right to control one's own body. Feminists fought for contraception and abortion rights in many countries where one or both were against the law, and for autonomy and equality within the couple. "The personal is political", proclaimed women inspired by Marxism and psychoanalysis. "Workers of the world, who washes your socks?" chanted demonstrators in the streets of Paris in the 1970s. In France, the Veil law legalizing abortion unleashed emotional debate in 1974.

Many Third World women could not identify with the struggles being waged in the West and insisted on leading their own battles at their own pace.However, these Western feminist movements breathed new life into the cause. Recognizing the changes and proclaiming their intention to accelerate them, the United Nations declared 1975 "International Women's Year" and organized the first international women's conference in Mexico City.

Already proclaimed in the Universal Declaration of Human Rights in 1948, sexual equality was reasserted in 1929 by the Convention on the Abolition of All Forms of Discrimination Against Women, which became a precious emancipation tool in the North as well as the South. At UN conferences in Copenhagen in 1980, Nairobi in 1985 and Beijing in 1995, women from both hemispheres found common ground, demanding the right to "have a child if I want it, when I want it," rejecting Malthusian principles and claiming their place in political bodies that until then had decided the world's future without them, struggling against religious fundamentalism that jeopardized their modest gains.

Misogyny of the Political Class

Of course, the struggle of Kuwaiti women against those who have denied that the right to vote or Indian women against the forced abortion of female foetuses is not the same as American women's battle against their own fundamentalists or French women's campaign against the misogyny take different approaches depending on the continent and do not necessarily have the same priorities, but the struggle has nonetheless become global during the past several decades. In the last twenty-five years, women have gradually increased their presence in public life, although it can hardly be said that the doors are wide open for them. From Africa to Asia, women's organisations have multiplied and acquired experience.

But their victories remain incomplete and the future is uncertain. From the nightmare of Afghan women to the ways, in which equality is resisted in the so-called most advanced countries, the obstacles show that there is still a long way to go. Will women see the end of the struggle in this century that has just begun, the one, which supposedly belongs to them?

15

High World Trade Growth Vs. Output:
WTO sees Link to Globalisation

World trade in merchandise goods is expected to increase in volume by 8 per cent in 1995 down marginally on the very high 9½ per cent for 1994. Although the current outlook is for a further modest slowing next year, trade growth will remain above the average of the past decade.

Recent trade growth figures continue to exceed world production growth by a large margin in 1995 probably by a factor of almost three and next year close to double. This persistent pattern relates closely to the "globalisation" of the world economy; a process which, brings far-reaching benefits and which can be promoted through the further development of the multilateral trading system.

The recent growth is as follows:

- a 13 per cent rise pushed the value of world merchandise trade past the $4,000 billion mark for the first time, to $4,090 billion;
- an 8 per cent increase in the value of trade in commercial services, to $1,100 billion, after near stagnation in 1993;
- a 23 per cent increase in the dollar value of merchandise trade in the first six months of 1995 which, allowing for the depreciation of the US dollar,

> is consistent with a full-year growth in trade volume of 8 per cent.

Globalisation

Over the period from 1950 (when the process of trade libealisation through the early GATT Round got under-way) to 1994, the volume of world merchandise trade increased at an annual rate of slightly more than 6 per cent and world output by close to 4 per cent. Thus, during those 45 years world merchandise trade multiplied 14 times and output 5½ times. However, the excess of trade growth over output growth varied; from an average of a mere half percentage point in the period 1974-84 to nearly 3½ percentage points in the most recent 10 years. In fact, the excess during the years since 1990 has been much higher still but it is not yet clear whether or not this represents a permanent shift to a faster rate of increase in the world's trade-to-output ratio.

To the question "will globalisation continue?" In this regard one has to observe two factors—technological change and the evolving strategies of firms and individual investors—impart a natural momentum to global integration. It is government policies which can speed-up, slow down or even reverse progress on global integration. In this context, the role of non-discrimination—in particular, through the "most-favoured-nation" (MFN) clause—is examined.

The MFN Clause

MFN was the centrepiece of a multiplicity of bilateral trade agreements reached in Europe in the second half of the 19th century, a period marked by very low tariffs and rapidly increasing trade. In contrast, the 1920s and the 1930s saw effects to restore liberal trade through international trade conferences rather than legally-binding commercial treaties based on MFN. The failure of these efforts contributed to the Great Depression and provided some of the roots of military confrontation in 1939. It was only after the War that negotiations established what became the GATT, a multilateral contract consisting of rules and disciplines and based firmly (Article) on MFN treatment.

The GATT system has been a post-war bulwark against a return to the trade chaos of the 1930's. In the 1990's, a disintegration of the globalised international economy on the scale of 1930's is almost unthinkable. In contrast, today "the threat that would be posed by a loss of credibility of the multilateral rules" (now represented by the WTO) would be "a fracturing of the global economy into inward-looking and potentially antagonistic trading blocks".

One can suggest two safeguards against such an eventually:

- the examination of new ways to ensure that free-trade areas and customs unions remain outward-looking and complement rather than compete with the multilateral trading system; and
- progress in dealing, at the multilateral level, with new issues tied directly to the further evolution of the global economy. These include telecommunications, financial services, environment, competition and investment policies among others.

Progress in dealing with these and other issues at the multilateral level will have a significant impact on the future pace of global integration, both directly and through its impact on the credibility of the multilateral system in influencing the broad spectrum of national trade policies.

16

Rural Poverty in India

"It is morning in a remote farming area in India. As her husband harnesses a bullock to plough their field, a women pounds the gain she will use for the day's main meal. Three kilometres away, their children are collecting fuel wood and water before starting their morning walk to school".

"After School, they help their mother light a fire with a few sticks, milk the cow and collect the sundried grain. That evening, as the family rests around the hearth, father worries about how to sell his onions before they spoil and the price falls. Before sleeping his wife prepares a basket of home-grown vegetables to sell next day at the village market five kilometres away. With the takings, she hopes to buy a kerosene lamp although she might not have enough cash left to buy the kerosene immediately..."

That description of rural life is a daily reality for hundreds of millions of families throughout India. Rural poverty, 1990s means subsistence on the meagre earnings of wage labour or unreliable harvests from small plots of land. It means raising a family without safe drinking water or proper sanitation, suffering disease or injury without medical assistance. In times of unemployment or crop failure, it means living with the pangs of hunger—and the risk of death by famine.

Inside the Poverty Trap

Poverty in rural India is created and perpetuated by a number of closely interlined socio-economic processes.

1. Policies and institutional arrangements biased against the poor exclude them from the benefits of development, frustrate their productive potential and accentuate the impact of other poverty processes.

Institutional processes that perpetuate rural poverty include lack of access to land, inequitable share-cropping and tenancy arrangements, poor markets, limited access to credit, inputs and technology, and ineffective extension services. Other constraints are lack of training facilities, inadequate research related to smallholder farming systems, and last but not least a lack of gassroots institutions needed to foster people's participation.

Policy and institutional biases have short and long-term impact. In the short term, the poor are unable to earn enough to meet nutritional requirements or to take advantage of the market. "In the longer term", "poor households continue to lag behind because they do not generate a surplus for investment, nor do they have access to investment opportunities. Moreover, the rural poor may be forced to overuse resources, which undermines productivity and income".

2. Even today dualistic agrarian structures originating in colonial times persist. In Indian, highly capitalised large and medium-sized farms have virtually monopolistic control over land and labour at the expense of the small farm sector. Large scale commercial producers—control the best farm land. Resources have been funnelled into irrigated plantations producing cotton and mechanized cultivation of sorghum. In marginal areas, mechanisation has led to environmental degradation and the loss of seasonal grazing and stock routes for pastoralists.

"Thus, side by side with modern agriculture, millions of marginal farmers and herdsmen subsist far below the poverty line". This dualism severely limits their capacity to grow food and accumulate capital. They lack marketable surpluses, and incentives and opportunities to save and invest.

3. Rapid population growth can cause and perpetuate rural poverty by increasing pressure on limited

productive resources, social services and employment, as well as paradoxically-creating labour shortage through outmigration.

The most obvious consequence of rapid population growth is that, even with relatively high rates of economic growth, improvements in living conditions are limited. Total saving in the economy declines, leaving fewer resources for investment in human development. Negative consequences are most acute in rural areas. Growing population often combined with traditional laws of inheritance—has led to fragmentation of holdings, degradation of crop and pasture land, and falling yields. In areas with unequal distribution of land, rapid population growth has accelerated proletarisation of the rural work force and reduced incomes.

4. Rural poverty malnutrition and undernutrition are closely linked to environmental degradation. Poor people in marginal areas are destroying natural resources as they struggle to keep their production systems sustainable. In acute shortage of arable land has forced farmers to reduce the length of fallow periods and plough up land previously reserved for grazing. These practices have led to declining yields, soil depletion and further impoverishment. Population pressure is pushing weaker members of the rural community into ecologically vulnerable areas.

Degradation of the environment is strongly linked to household food insecurity and lack of fuel. Much of the fragile forest cover has been destroyed by poor rural people in the search for grazing land and fuel-wood.

Government policies have also wrought environmental damage. A rapid expansion of areas under crops often accelerates deforestation and land degradation. Programmes to expand cereal production into marginal areas, subsidized capital to support commercial operations subsidies for inappropriate technologies and excessive transfer of income out of the agricultural sector may undermine the sustainability of smalholders and pastoralists' production systems.

Inadequate public investment in off-farm employment and infrastructure, a lack of price incentives and inadequate access to modern agricultural inputs and services discourage investment in land conservation, leading to further overuse and degradation.

5. As poverty undermines traditional social bond, the marginalisation of women has become a fact of rural life in India. With little or no access to land, millions of women depend on casual employment on meagre wages. Often, they farm fragmented plots of poor quality. Limited access to inputs, extension, training and credit limits, in turn, their ability to enter commercial agriculture.

The exodus of males in search of work in urban areas (itself an indicator of poverty) has serious consequences for the women they leave behind. Output from land often falls and less attention is paid to maintenance, setting the stage for a long-term decline in productivity. Many female headed households have abandoned the use of oxen for ploughing, some plough and plan late and others no longer weed their fields.

6. The ethnic or cultural marginalisation of tribal or minority populations also plays a role in poverty. Many of these groups are further threatened by newly marginalized groups—as the expansion of cultivation reduces the grazing areas of nomadic herders.

7. Exploitative middlemen also perpetuate rural poverty. Landlords exploit share croppers and tenants, moneylenders exploit debtors, and traders exploit small scale producers. During seasonal food shortages, the poor may have to borrow money at interests rates exceeding 20% a month. Force to devote most of their energies to debt servicing, they sink deeper into the poverty trap.

In some cases, government controlled co-operatives and government agencies whose task is to protect the poor may themselves practise forms of exploitation. Heavy levies imposed by government agencies have damaged small farmers. Large, inefficient bureaucracies are paid for by the productive sectors of the community and frequently contribute to the accumulation of large budget deficits.

8. Political troubles and civil strife have had a disastrous impact on the rural poor one effect is the disruption of development assistance to the rural poor, both from national and international agencies Another is the transformation of many producers into consumers of social services with serious consequences for production, savings, capital accumulation and investment.

9. The international economic environment directly influences the well-being of the Indian poor. Falling commodity prices and protectionist policies in India affect the employment and incomes of plantation workers and smallholders producing for export, particularly those relying heavily on a few agricultural commodities. Changes in international interest rates have repeatedly hurt small-scale producers in debt-burdened India, While world grain price increases has triggered rural famines.

The net flow of development resources to agriculture also affects rural poverty. Official development funding for food and agriculture increased between 1975 and 1982, but has fluctuated irregularly since. Moreover, concern with trade balance is diverting resources to export crops, sometimes at the expense of traditional crops grown by poor farmers.

17

Development: *The Third Way*

While great claims are being made for the increasingly more efficient and effective technologies perfected to serve development of the people in this scientific age, huge problems are threatening the globe. The problems are mass poverty and hunger, underdevelopment, waste, unemployment, resource scarcity, environmental destruction and armed conflict.

In finding answers to the prevailing problems we must be clear about the meaning and purpose of development. The first glaring mistake made is that development is interpreted as development of the economy and not as the total development of society. When economic development is made the supreme goal, most of the other vital aspects get ignored, namely, development of the political system, community, social cohesion, the ecology, culture and values, development and stability go hand in hand while poverty and chaos constitute the antithetical twin.

Appropriate Development

The key elements in the conception of appropriate development consist of: first, aiming at sufficiently comfortable material living standards and not affluent standard of the rich as in prosperous nations. Second, development must not be confused with GNP growth. Mere increase of economic activity must not be pursued exclusively at the cost of articles that are urgently needed by the poor majority to maintain them at a reasonable level of material living. Third, in the villages, we must produce articles as are needed by the villagers. Fourth grassroots and participatory development is essential so that the local people identify and solve their local problems. Fifth, instead of capital and energy intensive high technology, use labour-intensive technology, Instead of heavy industrialisation, promote medium scale industries and technologies. And, sixth, instead of preoccupation with a high GNP growth rate, focus on the development of communities and of rural bodies and take care to conserve the local ecosystems. The main purpose should be to meet the primary needs of ordinary people and promotion of their productive resources such as land.

In developing countries like India where billions of poor people remain condemned by conventional economic development strategies and theories, it is vital to introduce appropriate development measures to remove deprivation and ensure the necessity of modest living standards.

The world has witnessed the operation of the two systems namely, the capitalist and the socialist one that have obtained in different countries. Although both the systems have underlined the welfare of all as the basic goal, both have left a legacy of waste, hunger and gross human inequality. Some 1000 million people do not get enough to eat including some 20 million in the USA.

Third World Way of Development

The iniquitous situation in the present day world has sparked off fierce controversies among the conventional economists and the new radical economists who champion a third way, as the alternative way to serve the primary goal of all humanity to have sufficient means to lead a comfortable peaceful life.

In order to achieve prosperity, conventional economists have emphasized the production of bigger cake on the assumption that everyone will get a slice of it. They also argue that a "tide will lift all the boats". Both these assumptions have proved false in that the poor have neither the slice of cake nor has their boat been lifted. Third way system lays stress on highly localized, less cash-reliant and simply structured set-up. It should not be dependent on transport of goods, but concretrate on more local production to meet local needs with a role for barter and free exchange. He urges a radical re-think of conventional economics.

Is this Stepping Backwards?

The most common criticism levelled against the Third Way is that it will arrest the progress made hitherto and that it might mean a return to a 'primitive' way of life. There should be no fear on this score because the Third Way aims at the reduction in the use of resources and therefore, of excessive production and consumption. It does not in any sense mean stepping backwards to a lower level of the quality of life. Nor is the alternative way intended to destroy capitalism or socialism. The conception of the alternate way is to promote economic growth compatible with capitalism and socialism. The ground idea is to promote selflessness, mutual concern and social responsibility. This will replace selfish, competitive and avaricious attitudes as have developed in the conventional economic order of today.

NGO's Resolution at the Rio Conference

That there is increasing awareness of the threat posed by the growing power of multinational corporations was articulated by the international NGO forum in its Declaration resolved on 12 June 1992 at the UN Conference of Environment and development in Rio de Janeiro. The declaration states that "the Bretton Woods institutions have served the major instruments by which the destruction policies have been imposed on the world" and calls upon "the world's people to protect their economic, social, cultural and environmental interests against the growing power of transnational capital". The declaration further avers "we recognize the central place of spiritual values and spiritual development... and values of simplicity, love, peace, and reverence for life.

After the failure of the socialist system over four decades to achieve prosperity for all, the Indian Government switched over to the global market economy and is steaming ahead with added liberalisation measures to attract foreign investment and the multinational corporations (MNCs). Some adverse effects of this are already visible: for example, majority financial equity granted to MNCs and the emergence of foreign subsidiaries with cent per cent financial equity; the introduction of pizza and Kentucky Fried Chicken which has been detested by the people in Karnataka. The farmers have also revolted because their rights to produce and sell seeds have been wrested by foreign MNCs who have acquired patent rights over certain Indian crop seeds. In this scenario, the third Way has much to commend itself to the Government. The Third Way has the air of the Gandhian model of economy and production which emphasizes production by people for their own needs and preference for small and medium sized industry. The same paradigm was championed by the renowned economist Schumachar when he said "Small is beautiful". India should take good care against the present-day headlong drive for the entry of foreign capital and foreign heavy industries.

18

Consuming the Future

Now that we are to reach six billion of us, it is a good point to check again on what sort of lifestyles we pursue and what is the environmental impact of those lifestyles. It is curious that we have spent several decades being concerned about the growing numbers of humankind while not giving at least an equal amount of attention to the levels of living we aspire to, and how many natural resources we chew up thereby and how much pollution and waste we cause.

Everybody is a consumer of sorts. True, every fifth person scarcely qualifies for that designation, consuming goods worth less than $1 per day. Conversely, every seventh person qualifies for a designation of super-consumer, with a cash income at least fifty times greater. These latter are the people who, through their carbon dioxide emissions, are disrupting everybody's climate dozens of times more than the average citizen of the One Earth. Fair Play, anyone?

Much as the have-nots seek to match the have's, it is plain their efforts will not work out for a long time to come, at best. If every Chinese person were to consume just one additional chicken per year and if the said chicken were to be raised primarily on grain, this would account for as much grain per year as all the grain exports of the number two exporter, Canada. If the Chinese were to raise their per-capita consumption of beef, now only 4 kgs per year, to that of Americans, 45 kg, and if the additional beef were produced largely in feedlots after the manner of the United States, it would account for as much extra grain as the entire US grain harvgest, less than one-third of which is exported. Because of its recent climbing up the food chain toward a meat-based diet, China has become one of the world's leading importers of grain. The global grain market today is around 200 million tons per year, and shows scant scope for significant increase.

As a further measure of its ambitions, the Chinese government has designated the auto industry as one of five industry "pillars". Today China has fewer cars than Los Angeles. If per-capita car ownership, together with oil consumption, were to match that of the United States, China would need 80 million barrels of oil per day—by contrast with the world's 1996 oil output of 64 million barrels of oil per day. The surge in carbon dioxide emissions would be unprecedented.

All this notwithstanding, there are already some 250 million newly affluent people in China. They are people with a household income equivalent to perhaps US$20,000, and enough discretionary income to enjoy the perquisites of the good life as perceived by these nouveaux riches. Top of the shopping lists are meat and more meat, followed by cars whether big or small. These are the badges of success: they show you have arrived.

The new consumers in China are matched by at least 200 million in India, and tens of millions in South Korea, Taiwan, Malaysia and Thailand (the recent economic setbacks have not permanently punctured the economic bubblies). Then there are 200 million more in Brazil, Argentina, Venezuela and Mexico, and more again in Hungary and other countries of Eastern Europe, also Turkey. Put them all together and they total about as many as the 800 million long established consumers in the ultra rich countries (the OECD grouping). When the current economic hiccups in Asia are left behind, the ranks of the new consumers can be expected to rise rapidly.

But they cannot hope to become super consumers. Where would all the extra gain come from? How could the global climate tolerate the huge additional pulse of carbon dioxide? There are all kinds of other environmental reasons to suppose that environmental constraints will become all the more constraining. True, technology could help moderate the environmental impact. We could enjoy twice as much material prosperity while using only half as much natural resources and causing half as much pollution and waste. But the new consumers will want to pursue the American dream to the hilt, and it is hard to see that the best technologies could enable huge numbers of affluent aspirants, perhaps two billion people by 2010, enjoying even half the material prosperity of Americans with average household incomes of $40,000.

But is it true "prosperity"—mental and emotional as well as material? Or is the American dream becoming a nightmare with its harried lifestyles and declining leisure time, where the shopping mall is the ultimate Mecca, and the good life is a case of piling up goodies?

In any case, we cannot expect the new consumers to forego their "rightful share" of affluence unless the long-time affluent agree to cut back on their environmental ruinous lifestyles. It is these communities that must offer a strong example, and soonest. Where is the political leader who will espouse the new vision, however much it may be perceived as the ultimate vote loser?

19

The Population Challenge

During the last half-century world population has more than doubled, climbing from 2.5 billion in 1950 to 5.9 billion in 1998. Those of us born before 1950 are members of the first generation to witness a doubling of world population. Stated otherwise, there has been more growth in population since 1950 than during the 4 million years since our early ancestors first stood upright.

This unprecedented surge in population combined with rising individual consumption, is pushing our claims on the planet beyond its natural limits. Water tables area falling on every continent as demand exceeds the sustainable yield of aquifers. Eventual aquifer depletion will bring irrigation cutbacks and shrinking harvests. Our growing appetite for seafood has pushed oceanic fisheries to their limits and beyond. Collapsing fisheries tell us we can go no further. The Earth's temperature is rising, promising changes in climate that we cannot even anticipate. We are triggering the greatest extinction of plant and animal species since the dinosaurs disappeared. As our numbers go up, their numbers go down.

These effects of population growth are relatively recent, but assertions that population growth could affect human welfare are not. In 1798 Thomas Malthus, a British clergyman and intellectual, warned in his famous piece, *An Essay on the Principles of population,* of the tendency for population to grow exponentially while food supply grew arithmetically. He saw a world where human numbers would continually press against available food supplies.

During the 200 years since Malthus issued his warning, famine has visited countries as diverse as Ireland and India, Ethiopia and China. Indeed, despite the near-tripling of the world grain harvest since 1950 the hungry and malnourished in 1998 number an estimated 840 million—nearly as many people as lived in the world when Malthus penned his essay.

But the nature of famine has changed. Whereas it was once geographically defined by areas of poor harvests, today famine is economically defined by low incomes in those segments of society that lack the purchasing power to buy enough food. Famine concentrated among the poor is less visible than the more traditional version but is no less real.

In addition to checks imposed by food shortages, there is evidence that other checks on population growth are now emerging, such as new infectious diseases, including AIDS, Ethnic conflicts within societies, such as Rwanada and the Sudan, are also taking a growing toll. Water shortages on a scale that would deprive people of enough water to produce food could undermine governments.

The evidence gathered here indicates that the rapid population growth prevailing in a majority of the world's countries is not going to continue much longer. Either countries will get their act together, shifting quickly to smaller families, or death rates will rise from one or more of the stresses just mentioned. As human demands press against more and more of the Earth's limits, the questions is not whether population growth will slow, but how. Will it be because countries do it humanely by shifting quickly to smaller families? Or because they fail to do so, and nature ruthlessly imposes its own constraints? In a world facing many challenges as it prepares to enter the next century, this may be the most challenging of all.

Estimates of future numbers are based on the latest United Nations population projections, using their medium level figures. Under this scenario, world population will grow from 6.1 billion in 2000 to 9.4 billion in 2050 a gain of 3.3 billion. The other two U.N. projections put global population in 2050 as high as 11.2 billion or as low as 7.7 billion. While the medium scenario is judged by the U.N. demographers as the one most likely to materialize, it is not an inevitable population part for the next century. Indeed, because the projections are based exclusively on demographic assumptions and do not take into account the environmental limits to carrying capacity, they should be viewed as a first pass rather than the final word on estimates of future population.

We use the medium-level projections to give an idea of the strain this "most likely" outcome would place on ecosystems and governments, and the urgent need to break from the business-as-usual scenario. The mid-level projected growth in population of 3.3 billion by 2050 is very close to the growth that will have occurred between 1950 and 2000, some 3.6 billion. But there is one difference. During the half-century now ending, the growth occurred in both industrial and developing countries. During the next half-century, the entire burden of the projected increase of 3.3 billion will be in developing countries, many of which are hard-pressed to satisfy even existing demands on resources. In fact, the population of the industrial world is expected to decline slightly.

The annual rate of world population growth reached its historical high in 1964 at 2.2 per cent. Since then, it has been slowly declining, dropping to 1.4 per cent in 1998. Despite the falling rate of growth the number of people aged each year increased from 72 million in 1964 to the all-time peak of 87 million in 1990. Since then the annual addition has also declined, falling to 80 million in 1997, where it is projected to remain for the next two decades before starting to decline.

The population projections for individual countries vary more widely than at any time in history. At mid-century populations were growing every where, but today they have stabilized in some 32 countries, while they continue to expand in some countries at 3 per cent or more a year, Indeed, the world can be divided demographically into two camps: countries that have achieved population stability or are well on the way to doing so, and those that have not.

With the exception of Japan, all the nations in the first camp are in Europe. And all are industrial countries, the populations of some countries, including Russia, Japan, and Germany, are actually projected to decline somewhat over the next half-century. In addition to the 32 countries, containing 12 per cent of world population, that have stabilized their populations, in another 39 countries fertility has dropped to replacement level (roughly two children per couple) or below. Among the countries in this category are China and the United States the first and third largest countries, which together contain 26 per cent of the world's people.

Although fertility in these 39 countries has fallen below replacement level, their populations have not yet stabilized because there is a disproportionately large number of young people moving into the reproductive age group. Thus even if they hold their fertility at replacement level, population may continue to grow for several decades before it stabilizes. It was this realisation that led China nearly 20 years ago to shift its goal from a two-child to a one-child family. Leaders in Beijing realized that, if they did not do this they would be faced with adding the equivalent of another India to their population—a development they considered potentially disastrous for their people.

In contrast to this group some countries are projected to triple their populations over the next half-century. For example, Ethiopia's current population of 62 million will more than triple, as it climbs to 213 million in 2050, Pakistan's population is projected to go from 148 million to 357 million, surpassing that of the United States before 2050 today to 339 million, giving it more people in 2050 than there were in all of Africa in 1950. From an environmental Vantage point, considering particularly the availability of water and cropland, it is unlikely that the projected population increases for these three countries, and other countries with similar projected gains, will materialise.

As hard as it is to imagine the addition of another 3.3 billion people to the world's population, it is even more difficult to understand the effects of adding such numbers. As we look back over the last half-century, we see that World lumber use more than doubled, paper use increased nearly sixfold, grain consumption nearly tripled, water use tripled, and fossil fuel burning increased some fourfold. The relative contribution of population growth and rising affluence to the growth in demand for various resources varies widely. With lumber use, most of the doubled use is accounted for by population growth. With paper, in contrast, rising affluence is primarily responsible for the growth in use.

One way to understand the consequences of future population growth is to contrast some of the key trend projected for the next half-century with those of the as one. For example, we have seen a new fivefold growth in the oceanic fish catch and a doubling in the supply available per person, but biologists now believe we may have "hit the wall" in oceanic fisheries and that the oceans cannot sustain a catch any larger than today's. Thus people born today are likely to see the catch per person cut in half during their lifetimes.

Grainland per person has been shrinking since midcentury, but the drop projected for the next 50 years means the world will have less grainland per person than India has today. Future population growth is likely to reduce this key number in many societies to the point where they will no longer be enable to feed themselves. Countries such as Ethiopia, India, Nigeria, and Pakistan will see grainland per person shrink by 2050 to less than one tenth of a hectare (one fourth of an are) far smaller than a typical suburban building lot in the United States.

Given that at the amount of fresh water produced each year is essentially fixed by nature, the water available, per person has shrunk steadily as a result of population growth, leading to severe water shortages in some areas. Countries now experiencing these shortages include China and India, along with scores of smaller ones. As irrigation water is diverted to industrial and residential uses.

The challenge to governments presented by continuing rapid population growth is not limited to natural resources. It also includes education, housing, and jobs. During the last half-century the world has fallen further and further behind in creating jobs, leading to record levels of unemployment and under employment. Unfortunately over the next 50 years the number of entrants into the job market will be even greater. Few things threaten the political stability of a country as much as growing as growing ranks of unemployed young people.

As noted earlier, the U.N. population projections cited here are based on exclusively demographic assumptions, which are not related to the population carrying capacity of local eco systems. These projections are purely statistical, based on historical data on fertility, mortality, and average life span and assumptions about future trends.

Based on the analysis in it, I conclude that the medium projection of 9.4 billion people in 2050 which U.N. demographers consider to be the most problem is unlikely to materialize. Rather the world is more likely to follow a path closer to the low population projection of 7.7 billion by mid-century.

What is less clear is whether we will move to the lower trajectory because countries with rapid pollution growth quickly shift to smaller families or because they fail to do so and the resulting inability to manage threats from disease, spreading hunger, or social disintegration leads to rising death rates.

20

Population Growth and Jobs

Since mid-century, the world's labour force has more than doubled—from 1.2 billion people to 2.7 billion, outstripping the growth in job creation. As a result, the United Nations International Labour Organisation estimates that nearly 1 billion people, approximately 30 per cent of the global work force, are unemployed or underemployed (working but not earning enough to meet basic needs). Over the next half-century, the world will need to create more than 1.9 billion jobs—all of them in the developing world—just to maintain current levels of employment.

As economists often note, while population growth may boost labour demand (through economic activity and demand for goods), it will most definitely boost labour supply. During the next 50 years, almost 40 million people will enter the global labour force—defined as those between the ages of 15 and 65 seeking work—each year. Between 1995 and 2050, some 1.9 billion additional jobs will need to be created to absorb these new would-be workers. The most pressing needs will be found in the world's poorest nations—a sobering example of the vicious cycle linking poverty and population growth.

As the children of today represent the workers of tomorrow, the interaction between population growth and jobs is most acute in nations with young populations. Nations such as Peru, Mexico, Indonesia, and Zambia with more than half their population below the age of 25 will feel the burden of this labour flood. In the Middle East and Africa, 40 per cent of the population is under the age of 15. Since new entrants into the labour force were born at least 15 years ago, measures to reduce population growth have a delayed effect on the growth of the labour force, highlighting the urgency of taking action on population.

Nowhere is the employment challenge grater than in Africa, where at least 40 per cent of the population lives in absolute poverty. Although 8 million people entered the sub-Saharan work force in 1997, by 2030 this resource-scarce region will have to absorb more than 17 million new entrants each year. Over the next half-century, Nigeria's labour force is projected to grow by 246 per cent and Ethiopia's will soar by 337 per cent—both faster than growth of the general population. At current growth rates, the size of the labour force in sub-Saharan Africa will more than triple by 2050.

As a result of unprecedented population growth and increasing acceptance of female participation in the work force, the number of people seeking jobs in the Middle East and North Africa, a region already plagued by double-digit unemployment rates, will double in the next 50 years. In Algeria, where unemployment stands at 22 per cent, the labour force is growing at a staggering 4.2 per cent annually, and the number seeking work will more than double by 2050. Egypt alone will need to create 26 million more jobs by 2050 as its total population hits 115 million.

Nations throughout Asia will also see phenomenal increases in the numbers seeking work, including Pakistan, where the work force will grow from 70 million in 1998 to 205 million by 2050. Over the next 25 years, India will add nearly 10 million to its work force each year. During the same period, China will add nearly 6 million annually due to population growth alone, compounding the work shortages caused by the current flood of migrants to China's coastal cities and by massive layoffs—estimated at more than 30 million—as state-run operations are scaled back.

Nations are hard-pressed to educate and train rapidly growing numbers of young people in marketable skills for the global workplace. Moreover, meeting the basic needs of a growing population draws scarce foreign exchange and other resources from investments in education and job creation. Throughout the world, young people entering the work force are increasingly faced with unemployment and social marginalisation. In most societies, unemployment rates for those under 25 are substantially higher than for older people.

Surplus farmland once served as a traditional source of employment for growing populations, as new land could be ploughed to generate work and income. However, global percapita Greenland has dropped by half and considerably more in certain nations since 1950. Moreover, the machanisation of agriculture fuels the exodus of job seekers into the world's urban areas, where unemployment is often most acute, heavily reliant on natural capital in the past, future job creation will require massive amounts of financial capital to jump-start the industrial and service sectors.

As the balance between the demand and supply of labour is tipped by population growth, wages—the price of labour-tend to decrease. And in a situation of labour surplus, the quality of jobs may not improve as fast for workers will settle for longer hours, fewer benefits and less control over work activities.

Employment is the key to obtaining food, housing, health services, and education, in addition to providing self-respect and self-fulfilment. Rising numbers of unemployed people could drive global poverty and hunger to precarious levels, fueling political instability.

21

Opening Markets for Agriculture

While the Uruguay Round made a good start—more was done to liberalize agricultural trade and to bring agriculture into the system than in all previous rounds combined—we have to recognise that agriculture still has a long way to go to complete its reform and to be fully integrated into the world trading system. Prior to the Uruguay Round, agricultural trading rules were not in concert with other sectors. The Uruguay Round Agreement (URAA) made good first steps toward bringing agriculture into conformity with international trade rules governing other goods, but much remains to be done.

The Uruguay Round, of course, required certain reductions in trade-distorting measures, and the implementation of those reforms has proceeded very well. Two other legacies of the Uruguay Round are very important for the new negotiations—a mandate to continue what was begun, and a structure for achieving liberalisation. The WTO's "built-in" agenda includes agriculture. It was recognized from the outset that the first period of reform that we are still implementing was only a down payment.

In addition to the commitment to continue negotiations, the URAA—focusing on export subsidies, market access, and domestic support—established a structure on which to build. Establishing a three-pillar structure was the most time-consuming undertaking in the round. Fortunately, we do not need to reinvent that wheel. The structure of the rules provides a logical approach for the negotiations, one which most seem to agree we should keep and build on.

Export Competition

Export subsidies are an illegitimate policy instrument, a symptom of a systemic imbalance in a nation's agricultural policies, the costs of which are borne by others. The costs of domestic policy choices should be borne by the country that chooses them, not foisted into its trading partners by subsidizing export. The Uruguay Round made a start at eliminating agricultural export subsidies: 36 per cent reduction of budget expenditures on export subsidies and 21 per cent reduction of quantities over a six-year implementation period. With experience to show that markets adapt, we should now be able to improve the pace of export subsidy reductions and eliminate the export subsidy scourge from agricultural trade. Export subsidies are not allowed in the WTO rules for any other industry. Their use constitutes a source of trade distortion and degradation to the environment, and there is no valid reason to keep them any longer.

Market Access

The Uruguay Roud progress on market access leaves much to be done. It left tariffs too high, and it did not create much new market access. The average non-agricultural tariff is now 4 percent, while the average agricultural tariff is over 40 percent, and tariffs on some products exceed 300 per cent. With a few exceptions, nontariff barriers were converted to tariff, and members were required to open up at least a small minimum access—3 per cent of domestic consumption initially, growing to 5 per cent by the end of the adjustment period—under tariff-rate quotas.

The stage has been set for real reform. Let access continue to grow and let all tariffs be reduced to a negotiated maximum level by the end of the transition period. In addition, an examination of the administration of tariff-rate quotas should lead to transparent and open systems.

Many WTO members note that importers were required to change nontariff barriers to tariffs and grant access, while no reciprocal disciplines were imposed on export restraints of exporting countries. Net food importing countries should be able to expect that if they open their border to international market, those international markets will deliver supplies as reliably to importers as to the domestic markets of exporters. Willingness on the part of leading exporting members to discipline export controls will reassure "food security" countries that expanding market access is not risky.

Domestic Support

The Aggregate Measure of Support was a success as a component of the Agreement on Agriculture and the insistence on reducing trade-disorient measures. The drive toward decoupled support ("green box") is the key. By the end of 1996, the United States had largely decoupled farm programmes so that payments to farmers were not linked to a requirement to produce. Other WTO members will also succeed in orienting their policies toward market signals. In the new round, further review and decreases in the aggregate measure of support will clearly lead to market-based agricultural trade.

A new buzzword that some countries are using to justify domestic support is "multi functionality" It is a buzzword for what everybody in agriculture has known for thousands of years: agriculture serves other purposes besides producing food and fibre. But the real problem with the discussion of multi-functionality is not semantic. It is the confusion between policy goals and policy instruments. If the United State appear skeptical about the implications of multi-functionality for WTO rules, the U.S. objection is not multi-functionality as a factual matter. Each country chooses social objectives for themselves. There is no inherent connection between those objectives and trade distorting agricultural policies.

New Issues

While the Uruguay Round established effective disciplines ink traditional problem areas, such disciplines have not yet been established in some new areas. As monopolies, state trading enterprises (STEs) can distort trade, and they frequently operate behind a veil of secrecy. The agricultural trading system has much to gain from WTO disciplines on STEs because they allow some countries to undercut exports based on open market transactions and restrict imports.

Biotechnology holds tremendous promise globally for food consumers, producers, and the environment. With the world's population growing by about 2 per cent annually, there are 80 million more mouths to feed each year. Some countries threaten to adopt policies regarding the importation and planting of bio-engineered crops and the labeling of products containing bio-engineered foods that are not based on scientifically justified principles. If our farmers are to meet the challenge of feeding an ever-increasing population with a sustainable agricultural system, then they must have access to the new bio-engineered varieties. We need to think about how the WTO can help facilitate this new technology.

Developing Countries

One of the critical components to a successful new round of negotiations will be the full participation of a substantially increased number of developing countries. Open trade in agriculture relieves farmers in developing countries of the burden imposed by protectionism and export subsidies, while reducing hunger and offering reliable supplies of food at reasonable prices.

22

Market Access:
Eliminating Barriers that Impede Trade

Market access for particular commodities to be restricted by high tariffs and tariff-rate quotas (TRQs). The administration of TRQ systems in different countries can impede and distort commercial decision-making. One of the most important accomplishments of the Uruguay Round agreement was bringing agriculture more fully under General Agreement on Tariffs and Trade disciplines. A principal implication of this is that trade in agricultural products can now be restricted only by tariffs-quotas, discriminatory licensing, and other nontariff measures are forbidden: Also, all agricultural tariffs were "bound" in the World Trade Organisation (WTO); tariff rates above a binding violate WTO obligations.

While creating a "tariff-only" system for agricultural products is an important advance, too many market access barriers continue to impede international trade of food and fibre products. Market access barriers deny efficient producers the opportunity to compete in other markets and limit the variety and quality of products available to consumer. Reducing and removing these barriers will be an important element of the WTO negotiations.

What Are the Issues?

Eliminating nontariff measures was a necessary first step to removing trade barriers, but many of the tariffs in place are still prohibitively high. For example, while the average trariff assessed by the United States on agricultural products is less than 5 per cent (and for industrial products is less than 5 per cent (and for industrial products is nearly zero), the average agricultural tariff assessed by WTO members exceeds 50 per cent.

Moreover, in some case, market access for a particular product is restricted to a tariff-rate quota (TRQ). Under a TRQ system, import opportnities are established for a specific quantity of imports at a low tariff. All other imports of a product are subject to high tariffs. All tariffs, including in-quota and out-of-quota tariffs, are now bound against increase and subject to further reductions, a situation that will be a top priority in the next round of negotiations.

Where TRQs remain as a transitional step before more open trade is achieved, further reform needs to be undertaken in the upcoming negotiations. In the Uruguay Round, countries generally agreed to open TRQs to allow imports equal to current levels of trade or where imports had been low, new access opportunities were established. Recent experience also indicates that the administration of the TRQ systems in different countries can impede trade and distort commercial decision-making. It is expected that these elements will be subject to further disciplines in the upcoming negotiations.

Similarly, we need to closely examine the rules for state trading import monopolies in agriculture. Use of these state traders may have been justifiable when more restrictions were allowed on farm trade, but in the tariff only regime. It is difficult to see why a government needs to insert itself between thrust of WTO principles, countries should use the upcoming negotiations to increase responses to market forces competition and transparency where single-desk buyers or other restrictions on the right to import exist.

Although the WTO has moved agriculture to a tariff-only system, in too many cases countries operate variable tariff systems that result in confusing and unpredictable tariff collection. Measures such a reference-price schemes, price-brand systems, and variable tariffs operating under a high WTO binding make it hard for businesses to know exactly what tariff they will have to pay when their product arrives at customs. The uncertainty and lack of transparency chills trade and leaves the systems open to potential fraud and abuse. In some cases, reference-price systems can disadvantage suppliers of products with particular grades or quality. Countries are likely to investigate the operation of such tariff systems in the upcoming negotiations.

One of the elements of the Uruguay Round agreement was to establish a special agricultural safeguard mechanism to protect particularly sensitive products against a flood of imports or to guard against a sudden drop in the price of imports. The agreement establishes specific criteria for triggering the safeguard mechanism. Countries are expected to review the operation of the safeguard and review whether to continue its use in the upcoming negotiations.

23

Richer or Poorer?

Achievements and Challenges of Ethical Trade

Ethical trade as an approach to supply chain management has mushroomed in recent years. Northern companies are becoming increasingly concerned with the 'ethics' of their operations and risks to reputation and productivity posed by bad employment practices in global supply chains. But can voluntary private sector codes really improve employment conditions in supply chains?

Ethical trade is one dimension of corporate social responsibility, bringing social issues into the mainstream of commercial supply chain management through the use of codes of conduct. It is sometimes confused with fair-trade which addresses terms of trading for smaller producers, and fosters greater responsibility in supply chain relations.

Ethical trade, on the other hand, focuses on workplace issues, requiring that supplier's in particular meet minimum employment, worker welfare and aspects of human rights standards.

Similar management systems are well established for product safety and environmental issues, Here we focus on the social dimensions of ethical trade and its codes of conduct yet the separation of social and environmental standards is increasingly artificial in global sourcing agreements. A plethora of codes are on offer. The most numerous are in-houses codes such as Nike's 233 company codes counted in 1999 and the figure is rising.

Suppliers have to comply with and pay for multitude of similar but different codes. Harmonising codes or establishing equivalence is on the agenda but has not yet halted the problem of 'code overload'

At a broader level, industry-specific codes have also been developed. The US Apparel Industry Partnership/Fair Labour Agreement adopted by a number of leading US merchandising companies is a good example. Industry standards are not new, as ISO and EMAS environmental management systems show. Building on ISO principles, Social Accountability International (formerly CEPAA) has development SA8000. This is an independent social standard that can be used as an auditable code throughout the private sector.

Ethical trade is partly a response to consumer and campaigning group pressure in globalised economy. Alliances of companies, NGOs, trade. Developing codes of conduct through a multi stakeholder approach is a striking aspect of ethical trade, bringing together companies, NGOs, trade unions and some government departments. An example of this collaborative approach is the Ethical Trading Initiative (ETI) in the UK. The ETI's baseline code of conduct that corporate members from various industries must comply with as a minimum standard is more than just a code, ETI aims to provide a learning environment and sponsors pilot projects in developing countries to test different methods of monitoring and verification.

Codes of conduct need to be assessed in term of content, implantation and impact. A number of professional auditing companies have moved into his area, some accredited to audit specific codes such as FLA or SA8000. Suppliers audited against a specific code undergo an inspection, and where non-compliance is found, have to take remedial action or risk failing the audit.

Social auditing is a complex process, however, and it can be difficult to spot work place abuse, such as sexual harassment or force overtime, Workers have little confidence in a process that appears to be linked with management, and fear that reporting issues could risk their jobs. Advocates of the multi stakeholder approach argue that effective monitoring and verification of codes must involve local NGOs and trade unions in which workers have trust. Participatory social auditing also a means of raising awareness and of facilitating behavioral change, can help reveal serious management problems. But in many developing countries local organisations lack the capacity to participate: developing sustainable local systems of monitoring and verification remains an important challenge.

Do the advantages of multi-stakeholder approaches outweigh immediate constraints? Ethical trade is a largely northern driven process, reflecting Western ethical thinking and priorities, Southern based initiatives, however, are expanding, raising the possibility of local ownership of codes, collaboration poses challenges. Stronger relationships and better understanding are essential between southern and northern workers, producers, trade unions, and NGOs for codes to work globally.

But there is still skepticism as to the extent of the benefits that ethical trade might bring. Will increasing southern capacity to participate, as the ETI has done in its pilot project, help? Will building trust, confidence and dialogue achieve the objectives of ethical trade, north and south? Child labour is often more complex, however, than codes make it appear. Codes need to address the conditions of all workers within the supply chain, including the least visible; partnerships must include all groups to address these limitations.

The role of government is hotly contested. Can a system whose credibility depends on under-resourced civil society actors, often excluding democratically elected representatives, maintain genuine credibility? If the boundaries between private sector and public sector roles are not defined, the list of private sector responsibilities will become unmanageable. Private sector initiatives are not a substitute for more comprehensive national or international development policies.

What are the consequences of codes? Do they encourage downsizing or reinforce from large suppliers where compliance is more easily monitored? There is a risk that the gains of some will be at the expense of others.

Ethical trade has successfully begun forging partnerships to find solutions. While it might be wrong to assume that ethical trade can change the world, handled wisely it could make a world of difference for some. Yet it is not a panacea for development. Issues that remain unchallenged by ethical trade include:

- The exclusion of companies producing for domestic markets—often bigger employers.
- Underlying causes of poverty and social marginalisation.

24

Tapping the Market:

Can Private Enterprise Supply Water to the Poor?

Over 170 million people have no access to clean water in urban areas through out the world. Inefficient operation of state owned water companies is at the root of this injustice: gross over-staffing and political interference in tariff-setting have starved utilities of the resources needed to expand piped networks to impoverished areas.

The failure of the supply-driven approach has led to public private partnership (PPPs) designed shift water utilities towards a demand-driven approach. Have these changes been accompanied by improved access to clean, affordable water for the urban poor? Has PPP improved equity in urban water supply? Are the new private sector operators addressing the needs of the urban poor in practice? This article examines the extent to which the urban poor have benefited or not from this newly emerging institutional arrangement.

The 'public' approach typically provides unclean water sporadically. It requires expensive, highly educated professionals, significant subsidies and tends to service clients on high and middle incomes whilst changing low tariffs. International financial institutions have failed to enable the public water suppliers to improve performance, either through massive investment in engineering, or through capacity building and institutional development.

At the other extreme is an efficient, demand driven, customer-oriented approach, the 'small scale independent providers', delivering water to cities' inhabitants, with near 100 per cent bill collection efficiency. Promoting local employment and servicing the poor, this approach has, until recently, been ignored by water sector professionals. Lacking regulatory oversight, however, their prices are typically 10 to 20 times higher than those paid by high-income consumers connected to the network. Which of these providers are most effective at serving the poor? The starting point is to recognise the evidence suggesting that the urban poor are prepared to pay to meet their survival and convenience needs for water.

Notwithstanding the rhetoric to the contrary by some trade unions and NGOs, initial results from larger cities indicate that the efficiency of 'privatised' water utilities has improved markedly: leakages are down and net revenue is up through improved billing and collection and reduction in personnel. Whether this is due to the alleged benefits of private sector investment or the freedom of foreign operators to manage without being beholden to employee and entrenched political interests is not yet clear.

Has the Extension of the Network to Poor Communities Been Speeded Up?

Concession contracts require private operators to meet coverage targets. But the decisions on the direction of network expansion to meet targets are usually left to the operators as regulatory bodies are usually formed after contract signing. So are poor communities given priority? Technical criteria based on cost-effectiveness in the construction of main pipes, commercial criteria based on pressure from property developers, and political criteria based on vote-winning tactics may all conflict with social criteria based on the pressing needs of poor communities.

Is the Cost of Household Connection Affordable by the Poor?

Even where operators do give priority to extending the piped network into poor communities, difficult issues arise over the financing of secondary pipelines, household connections and meter installation. Techniques are evolving to reduce the cost of connection so as to ensure affordability for all. These may take the form of tripartite arrangements whereby the public sector provides grants for the purchase of materials, community groups provide voluntary labour, and the private operator provides technical assistance. NGOs may contribute by providing crucial skills in team management and local understanding not usually found in the bureaucratic culture of public sector institutions or the technoprofessional culture of private water companies.

Is the Water Tariff Affordable by the Poor?

Even where poor communities have been connected, there is no assurance that householders can afford the water charges. Many have no job security or regular income. Billing arrangements need to be shortened from the monthly norm to fit the short-term financial horizon imposed by household poverty.

Property-based tariff structures still discriminate against the poor by providing far cheaper water per litre for high-income households consuming large volumes for swimming pools and sprinkler systems. Reforms should positively discriminate in favour of the poor, with some cross-subsidisation from richer to poorer households. However, where the initial life-line block is greater than average monthly domestic water use by the poor (perhaps over $6m^3$ per household a month), middle income groups benefits the most. A single volumetric tariff for domestic consumers with subsidies aimed at facilitating water connections rather than consumption is now being recommended.

This article focuses on the three major challenges for the sector in the new millennium.

- First it is crucial to develop regulatory skills to oversee this affordable expansion. The key capacity constraints facing municipalities—usually the public sector partners in PPPs.
- Second is the need to incorporate the skills of small-scale independent providers.
- Thirdly it is important to move beyond the metropolitan capitals, to where cross-subsidies are more achievable, and address the water needs of the urban poor in the myriad of secondary towns in the south.

25

Give Developing Countries a Morc Favourable Deal: *An Assessment of the World Trade Conference in Doha*

At the end of the 4th WTO Ministerial Conference in Doha, Qatar, the representatives of all WTO member states vigorously applauded Director-General Mike Moore when he dubbed the adopted work programmes for the new round of trade negotiations the "Doha development agenda."

The launching of a new of trade negotiations with a broad agenda was the objective persistently pursued by the industrial countries, in particular the European Union, the United States, Canada and Japan. This objective has been achieved. Besides the continuation of the negotiations in the fields of agriculture and services, the Ministerial Declaration adopted by the Conference provides for the opening of negotiations in eleven additional fields. Undoubtedly a success for the industrial countries.

Clear Mandate for a New Development Round

The negotiating mandate, though, clearly reflects the political will to make the new round a "development round" with the aim of significantly improving the integration of the developing countries into the world trading system. To a large extent it takes into account the specific interests of the developing countries. Certainly a success with which the developing countries can credit themselves. A crucial factor of the course and the successful outcome of the Ministerial

Conference was, without doubt, the active involvement of the developing countries in the preparatory and negotiating process.

Doha Determines Merely the Work Programme for Negotiations

The Ministerial Declaration adopted at the conference merely determines the work programme for the new round of trade negotiations. Three factors contributed decisively to the positive outcome of the Ministerial Conference. There was a broad consensus among the WTO members states that (*i*) a second Seattle-like failure would put the WTO's workability at risk and was to be avoided at all costs (the 3rd WTO Ministerial Conference I Seattle in December 1999 ended in chaos without the adoption of a Ministerial Declaration); (*ii.*) the recessionary trends in the world economy were to be countered with the successful conclusion of the Ministerial Conference in Doha to improve the prospect for short-term recovery and, thereafter, sustained economic growth; (*iii.*) in response to the terrorist attacks of September 11, 2001, there should be a clear commitment to strengthen the rules-based multilateral trading system. Failure was, therefore, not an option, The strategic conclusion drawn from the Seattle failure was to limit the Doha Ministerial Declaration to establishing a broad, generally-worded negotiating mandate for a new round of trade talks that does not anticipate the outcome of the negotiations on controversial issues. The strategy worked. The deliberations at the Ministerial Conference focused on the scope of the negotiating mandate. The task of reconciling the conflicting interest between industrial and developing countries and working out a fair compromise has been left to the forthcoming negotiations.

Recognition of the Interests of the Developing Countries

In view of the objectives of creating a basis for sustained economic growth in the developing countries by better integrating them into the world economy and increasing their share in world trade, important preliminary decisions with regard to the forthcoming negotiations were taken by the Ministerial Conference:

- the Ministerial Declaration stresses the importance of implementing and interpreting the Agreement on

Trade-Related Aspects of Intellectual Property Rights (TRIPS Agreement) in a manner supportive of public health and access to medicines; in recognition of the seriousness of the problem, a separate 'Declaration on the TRIPS Agreement and Public Health' was adopted; a number of public-health related issues have been referred to the Council for TRIPS for further deliberation;

- the Council for TRIPs has been asked to examine the relationship between *(i,)* the TRIPS Agreement and the Convention on Biological Diversity and *(ii,)* the protection of traditional knowledge, taking full account of the development dimensions;
- numerous problems regarding the implementation of WTO agreements are dealt with in a separate 'Decision on Implementation-Related Issues and Concerns' adopted by the Ministerial Conference; outstanding implementation issues are to be addressed as a matter of priority by the relevant WTO bodies;
- the Council for Trade in Goods will examine the proposal to bring forward the liberalisation of the textile sector under the Agreement on Textiles and Clothing;
- as regards agriculture, comprehensive negotiations were agreed on, aiming at: substantial improvements in market access; reductions of, with a view to phasing out, all forms of export subsidies; and substantial reductions in trade-distorting domestic support;
- as regards market access for non-agricultural goods, negotiations were agreed on, with the aim of reducing or, as appropriate, eliminating tariffs and non-tariff trade barriers in particular on products of export interest to developing countries;
- recognition of the principle of special and differential treatment of the developing countries as an integral part of all WTO agreements;

- technical cooperation and capacity building have been recognised in the Ministerial Declaration as 'core elements of the development dimension of the multilateral trading system' and firm commitments have been established in various paragraphs.

Turning the Ministerial Declaration's Spirit into Practical Policy

With these preliminary decisions regarding the agenda of the forthcoming negotiations, the course is set for the better integration of the developing countries into the world economy. To stay the course, there must be clear commitment and political will on the part of the industrial countries to make the new round a 'development round' by taking the developing countries' interest fully into account, being prepared to make meaningful concessions, and making good on the promise of significantly increased trade and investment-related technical assistance.

In the course of the negotiations it might prove a problem that many of the obligations in favour of the developing countries are formulated rather vaguely. The Ministerial Declaration is confined to declarations of intent even where—with a certain degree of goodwill—binding commitments would have been politically feasible. The bringing forward of the liberalisation of the textile sector, a key demand of the developing countries, has been referred to the Council for Trade in Goods for examination; this is certainly an expression of the industrial countries' willingness to compromise, but is in no way anticipates the final decision. As regards the objective of duty-free and quota-free access for all products of the least developed countries to the markets of the industrial countries the Ministerial Declaration simply repeats the commitment which was already expressed in the United Nations Millennium Declaration of September 2000, at the 3rd United Nations Conference on Least Developed Countries in Brussels in May 2001, and at the G7/8 Summit in Genoa in July 2001. Except for the European Union, no party has put this commitment into practice so far; the United States and Japan in particular have shown little enthusiasm for introducing duty-free and quota-free of all LDC products.

What makes us believe that the Doha Ministerial Declaration will make a difference? The chapter on agriculture is more specific in that it provides for negotiations aimed at significantly improved market access, reductions/phasing out of all forms of export subsidies, and substantial reductions in trade-distorting domestic support. However, a clear road map including a timetable for the negotiations and specific benchmarks for the reduction targets were beyond Doha's reach; in addition, the qualifier that the commitment to comprehensive negotiations does not prejudge the outcome of these negotiations leaves a back door open. In conclusion: If you remove the merely rhetorical phrases—such as "we place the developing countries' needs and interests at the heart of the World Programme adopted in this Declaration", "to take fully into account the development dimension...",—from the Ministerial Declaration, it becomes quite clear that the text contains relatively few 'programming elements' with a view to the development agenda of the forthcoming negotiations.

Fears that the vested interests of the industrial countries will re-gain precedence over development aspects in the course of the negotiating process are certainly not entirely baseless. The 'steel war' the United States is about to declare on the rest of the world clearly indicates that the Doha fair weather period is over. Business as usual has returned. The American steel tariff threats prompted EU Trade Commissioner Pascal Lamy to speak of a "perverse signal at a time when the ink is barely dry on the Doha Agreement."

The non-governmental organisations have a decisive role to play. It is their role to monitor the new round of trade negotiations, to make the negotiating process more transparent, to create public awareness with regard to the issues at stake, and to build up political pressure with the objective of making sure that development aspects are not pushed to one side and that the interest of the developing countries will makes their way into the agreements to be concluded.

Coherence of Trade Policy and Development Policy

The negotiating mandate for the new round of trade talks adopted in Doha has brought development politics onto the agenda of the WTO. The mention of development aspects in the WTO set of rules and regulations is not, in essence, new. In fact, the development dimension is recognized as an integral part of the general WTO mandate to foster economic growth. However, the particular importance the Doha Ministerial Declaration attaches to the consideration of development aspects in the negotiation process (it seeks, as it is put there, "to place the developing countries' needs and interests at the heart of the work programme") offers the opportunity to achieve greater coherence of trade policy and development policy. In this respect, the Doha Ministerial Declaration reflects the same trend as the "Everthing-but-Arms-Initiative" (EBA) of the European Union. Subsequent to its adoption by the EU member states, Pascal Lamy emphasized the coherence aspect as the characteristic feature of EBA Initiative (outweighing the shortcoming relating to bananas, rice, and sugar) by saying, "It is the first time that the European Union's trade policy has been substantially modified by the necessity of contributing to development policy." This perspective also characterized the 3rd United Nations Conference on Least Developed Countries in Brussels in May 2001.

To sum up, it can be said that the Doha conference has sent out an important signal for the process of coordinating trade and development policy with the long-term objective of achieving a coherent policy framework. The next step towards greater coherency can be taken at the International Conference on Financing for Development in Monterrey/Mexico in March 2002.

Sustainable Development as the Guideline for Further Developing the Multilateral Trading System

The Ministerial Declaration reaffirms the commitment to the objective of sustainable development, as stated in the preamble to the Marrakesh Agreement of April 1994 (i.e. the Agreement establishing the WTO). However, theory and practice are far apart. The negotiating mandate for the new round is too cautious a step towards integrating environmental and social aspects into the WTO set of rules and regulations to be able to bridge that gap. Looking at the three pillar of the sustainable development concept—economic development, environmental protection, and social protection, in a nutshell the following can be said:

The negotiating mandate for the new round deserves good grades as far as the first pillar, economic development, is concerned. The course is set for better integration of the developing countries into the multilateral trading system, thus giving them the chance of actually benefiting from further trade liberalisation in the form of trade-induced economic growth. The inclusion of the so-called 'Singapore issues', investment and competition, offers the prospect of a medium to long-term improvement of the business and investment climate in the developing countries. As for environmental protection; negotiations on a (very) limited scale have been agreed on, the desirability of further negotiations will be examined. This is certainly not a big breakthrough, but a first step towards integrating ecological aspects into the trade rules. Disappointingly (but not surprisingly), social issues were not dealt with at the Doha Ministerial Conference. The developing countries' resistance to even discussing social issues, such as core labour standards, in the framework of the WTO could not be overcome; the issue was considered an absolute 'deal-breaker'.

Outlook

The developing countries' consent to the launching of a new round of trade talks cannot disguise the fact that there are still significant differences of opinion over a number of issues, including such key issues as agriculture, environment, investment and competition, and that there is a great deal of mistrust on the part of the developing countries. The one-day extension of the Ministerial Conference alone is proof of how difficult the process of reaching consensus on the launching of a new round of trade talks and its agenda had been. In order to successfully conclude the new round, the industrial countries have to deliver on their commitments, such as improving market access for goods of export interest to the developing countries and increasing their trade-related technical assistance.

The assurance of increased technical assistance was a major bargaining chip in getting the development countries' OK for the new round. If insufficient funds for technical assistance and capacity building measures are provided, it will most certainly diminish the chance of getting quick results. In a comment on the forth coming negotiations, the British Economist also highlighted the credibility aspect and the need for significant concessions, "Poor countries remain deeply suspicious of the rich world's commitment to truly freer trade. They bitterly remember the Uruguay Round, whose benefits went mostly to the rich. For the new talks to succeed, those suspicions must be proven wrong. Europe and America must quickly open up their markets for farm products and textiles. They must show that environmental concerns are not going to become a backdoor excuse for renewed protectionism. They must reform their oft-abused system of anti-dumping rules. And they must deliver on promises to beef up poorer countries' capacity to deal with the intricate procedures in the world trading system."

The Doha Ministerial Declaration offers the prospect of long-term gains for the developing countries. However, turning potential into actual gains requires tenacity in pursuing policies aimed at improving the business climate and, in general, the framework conditions for economic growth. Increased trade-related technical assistance and improved market access will not automatically result in growing export volumes for the developing countries. In addition, the strengthening and diversification of productive capacity is required. Successful integration into the global economy depends on tackling the supply-side constrains and other 'behind-the border impediments to trade' (ranging from weak infrastructure. Insufficient ancillary services and poor governance to macroeconomic instability). The Tanzanian Trade Minister, Iddi Simba, emphasized the complexity of the problems the developing countries are facing in his statement at the Ministerial Conference: "To operationalise the development agenda we need to have adequate capacity building which will go beyond addressing the normal WTO obligations. Adequate resources in the form of financial and technology transfer need to be in place to address the supply-side constraints. Along the same lines, WTO Director-General Mike Moore stated, "Capacity problems [in producing goods and services competitively], not trade barriers, are the major obstacles to growth in developing countries."

Concluding Remark

By creating a rules-based multilateral trading system, the WTO set of rules and regulations contributes to the shaping of the process of globalisation and to the emerging system of global governance. However, it can hardly be disputed that so far the industrial countries have been the main beneficiaries of the WTO-driven economic globalisation. We are still miles away from a true win-win situation. In a recent interview with the German weekly *Die Zeit, the,* Nigerian President, Olusegun Obasajo, criticized the industrial countries ' countries hypocrisy, saying "Globalisation is a good thing. Bu only if there is a level playing field, from which all countries are able to benefit. You tell us that we have to open up our markets for your goods, whereas you keep your markets closed for our goods. Europe protects itself with innumerable trade barriers, everybody know that. What kind rules are those?"

That is exactly what matters. The new round of trade negotiations lunched in Doha must result in modified trade rules. Trade rules which take account of the specific economic constraints of the developing countries and are more favourable to them. The developing countries must be given the chance to 'cash in' on trade liberalisation and, strengthened by trade-induced economic growth, to pursue national pro-poor policies aimed at eradicating poverty.

26

Literacy Gaining too Slowly

In the past five decades, the number of adults who can read has increased by about 1.8 billion world wide—a net growth of some 1,20,000 people a day. This continuing improvement in global literacy from 56 per cent of the population in 1950 to about 74 per cent today represents encouraging progress. But is also hides troubling disparities between industrial and developing nations and between men and women.

In 1970, some 94 per cent of adults over age 15 were considered literate in industrial countries, compared with only 45 per cent in the Third World. Since then, literacy rates in the developing countries have gained ground impressively, climbing to 65 per cent. Unfortunately, that still leaves 1.4 billion illiterate adults worldwide. With population having grown even faster than literacy over most of the past tree decade, the absolute number of people who cannot read is greater now than it was in the early 1960s.

As education programmes proliferated in the developing world between 1995 and 2000, literacy finally began to gain on population growth, and the total number of illiterates' worldwide fell by 2.4 million–the first time the number has actually declined. At the rate of improvement, however, it would take 3,000 year for the number of illiterates to approach zero. Further more, as the income gap between rich and poor nations is now widening rather than narrowing, maintaining even this rate of improvement may be difficult.

The disparity between male and female literacy is pervasive, cutting across economic and regional lines. In 1970, about 70 per cent of the world's men were able to read, but only 54 per cent of the women. By 1990, both sexes had increased in overall literacy but the gap between the two had narrowed only slightly by 2 percentage points. At this rate, it would take more than 200 year for women to be as literate as men.

Even more disturbing is the fact that in some areas, the gender gap has actually widened. In Africa, while women's literacy climbed from 11 per cent in 1962 to about 30 per cent in 1985, the rate for men increased from 26 to 56 per cent during the same period; thus the gender gap increased from 15 percentage points to 26. Since 1985 however, African women have closed the gap slightly.

Illiteracy in industrial countries, while affecting a diminishing percentage of the population, appears to constitute a growing social problem. Because an increasing proportion of all jobs in these economies demand some facility reading and writing, those who lack these skills are often mired in chronic unemployment and poverty.

In the Third world, literacy is not only a key factor in economic productivity but is a basic—though often disregarded—factor in the vicious cycle wherein uncontrolled population growth hastens environmental degradation, leading to still more intractable poverty. Efforts to stabilize population growth are unlikely to succeed without fundamental improvements in the status of women-including their rights and access to education. Closing the literacy gap between men and women may thus be a significant measure of progress in human development.

Regionally, there are large difference in literacy. Both overall and in term of women's literacy, Africa is the most deprived of the continents. In 1990, less than a quarter of the adult population was literate in six countries in the world—all of them in Africa. In 19 nations, 17 of them in Africa, the lifetime total schooling averaged less than one year for each adult. And in Burkina Faso, Djibouti, and Somalia, where fewer than one woman in five could read, it averaged less than three months.

The functional importance of literacy to economic growth was formally recognized at a World Congress on Illiteracy in 1965, which gave impetus to the establishment of numerous national programmes. Subsequent statistical surveys demonstrate a high correlation between literacy rates and income. In real per capita income, the 20 wealthiest countries in the world in 1990 included 18 of the 20 nations with the highest adult literacy scores. The only notable exception was Ireland, which ranked near the top in literacy but had modest per capita income, albeit higher than in most developing countries.

Conversely, the greatest impoverishment is generally found in countries with lowest per capita gross domestic product (GDP) in 1990, the average adult had slightly more than one year of schooling.

These general observations are confirmed by studies showing that when literacy is increased in a given country, productivity also rises. A study of 88 countries found that a 20 to 30 per cent increase in literacy produced gains of 8 to 16 per cent in GDP. Another study concluded that, "about a fifth of income inequality could be explained by educational inequality."

Beyond economic motives, several other forces have propelled the movement that has led, during this century from a world in which a minority of people could read to one in which three of every four people can. At the 1985 International Conference on Adult education in Paris, literacy was identified as fundamental human "right to learn"

The idea of literacy as a means to liberation has spread widely, and has clearly been a factor in the global movement toward democracy.

With few exceptions, low literacy is associated not only with greater poverty but with greater likelihood of political and social instability—resulting in still further sapping of national assets as resources are directed toward military security rather than social investment. A telling measure of this phenomenon is the ratio of soldier to teachers, which is far higher in countries with low literacy and high poverty. It may be noteworthy that of 160 countries surveyed, the two with the highest soldier-teacher ratios were the every one the U.S. government perceived a need for military intervention in during the past three years; Somalia and Iraq.

27

The WTO and the Developing Countries

The special status of developing countries in the GATT will continue to receive recognition in the WTO. The preamble of the Agreement Establishing the WTO states that "there is a need for positive efforts designed to ensure that developing countries, and especially the least developed among them, secure a share in the growth of international trade commensurate with the needs of their economic development". In addition to retaining the provision that concerned developing countries in GATT 1947, the new agreement generally contain provisions for developing counties and least-developed countries, often consisting of longer transition periods for the full implementation of some obligations and various exemptions from obligations, particularly for the latter group of countries. Also, in some instances, the exports of developing countries benefit from a better treatment with respect to measures taken by other WTO Members. Technical assistance is to be provided to developing countries to assist them in assuming their obligations and more effectively realizing the benefits of the multilateral trading system.

Least-developed countries are singled out in the Final Act as requiring special attention. This is reflected in the agreements through a number of provisions which provide the most favourable treatment for this group in terms of rights as well as lower levels of obligations. In additions, the Decision on Measures in Favour of Least-Developed Countries, makes provision for measures of special assistance, including technical assistance "in the development, strengthening and diversification of their production and export bases including those of services, as well as in trade promotion, to enable them to mamimize the benefits from liberalized access to markets". As part of its functions, the Committee on Trade and Development (a subsidiary body of the General Council) will periodically review the special provisions in favour of least-developed countries and report to the General Council of the WTO for appropriate action.

The Declaration on the Contribution of the WTO to Achieving Greater Coherence in Global Economic Policy making identifies the need for strengthening the relationship between the activities of the WTO, the International Monetary Fund (IMF) and the World Bank as a way of ensuring greater coherence in global economic policy-making.

Market Access

Industrial Products. For developed countries, the main features of their market access commitments in industrial products include the expansion of bindings to cover 99 per cent of imports; the expansion of duty free access from 30 to 40 per cent of total imports; and the reduction of the trade-weighted average tariff by 40 per cent (i. e. from the per-Uruguay Round level of 6.2 per cent to the post-Uruguay Round level of 3.7 per cent). With respect to tariff reductions on individual product categories, developed countries will reduce tariffs by substantially above-average amounts (60 per cent or more) in three categories—wood, pulp, paper and furniture; metals; and non-electric machinery and reduce tariffs by less than the 40 per cent overall reduction in four categories—fish and fish products; textiles and clothing; leather, rubber, footwear; and transport equipment.

In terms of exports from developing to developed country markets, the total reduction in the average tariff of developed countries is 37 per cent. Below average tariff reductions, and above-average. levels of tariffs apply to labour-intensive manufacturers (textiles and clothing, leather goods) and certain processed primary products (fish products) that have been—and continue to be—regarded as "sensitive".

Developed countries constitute the most important merchandise exports markets for developing countries (62 per cent in 1992). In part, this is because the developed countries account for the bulk of global income and expenditures. At the same time, market access opportunities for developing countries in each other's markets have long been affected by the protection in their own market. In many instances, the level of protection is quite high, because important protection, ultimately, acts as a tax on exports as well, protection in developing countries also hinders the integration of developing countries also hinders the integration of developing economies, not just with each other, but with the larger global trading system.

The reductions in bound tariffs which the new commitments of developing economies represent are difficult to assess for several reasons. The first is that comprehensive information on base (1986) tariffs is unavailable in many cases as a result of the low level of bindings among developing countries. In these cases, the post-Uruguay Round average bound tariff usually involves a decrease in the ceiling bindings applied to items already found, combined with ceiling bindings above currently applied rates for previously unbound items. Another reason is that, where developing economies had bound all or a significant portion of tariffs prior to the end of the Round, the Uruguay Round tariff commitments often reflect a decline in ceiling rates (rather than applied rates).

At the same time, current tariff levels already reflect the often substantial reductions undertaken autonomously in the course of the Round. Though many developing countries may not be required to introduce further cuts, previous liberalisation has been at least partly locked in through new commitments on bindings. Ceiling bindings are considered to be so important that countries which agree to bind previously unbound tariffs are given "negotiating credits" for the decision even if the tariff is bound at a level above the currently applied level (as is the case for many developing economy participants in the Round). Bindings have also played a key role in establishing the domestic and international credibility of domestic reform programmes in many countries. Although an integral part of the tariff negotiations, bindings clearly are more akin to rules and procedures—in terms of their contribution to the predictability of future market access—than to direct increases in market access. However, even ceiling bindings yield significant benefits related to liberalisation when they reduce the expect value and variance of protection.

On the basis of data available for 26 developing countries, the GATT Secretariat has identified the main features of their market access commitments. These include: the expansion of bindings to cover 61 per cent of imports, compared to the pre-Uruguay Round level of 13 per cent. The increase in the security of trade among developing regions is reflected mainly in Latin America—where participants will bind 100 per cent of tariff lines at ceilings rates.

Too often, "market access" as used in descriptions of the Uruguay Round results is defined—implicitly or explicitly in a way that is too narrow and, even worse, mercantilist. The narrowness results from limiting the analysis of change in market access to changes in tariffs and quotas. This overlooks three other key aspects of market access, namely the bindings of tariffs, the rules and disciplines on the use of other trade-related government interventions, and the institutional arrangements for monitoring and enforcing compliance with those disciplines. It is progress in these latter three areas that determines the security of increases in market access from reductions in tariffs and the elimination of quantitative restrictions. Since the gains from trade liberalisation depend heavily on the stimulus it provides to traderelated investment, the security aspect is crucial.

Agricultural Products. Increased market access for agricultural products includes the "tariffication" of all non-tariff border measures (conversion to tariff-equivalents)—with the exception of those products for which special treatment has been negotiated—and a binding of all tariffs on agricultural products. As a result, the security of trade in agricultural products will for the first time be greater than in industrial products, since 100 per cent of agricultural product tarifflines will be bound.

Tariffs resulting from the "tariffication" process, together with the other tariffs on agricultural products, are to be reduced by a simple average of 36 per cent over six years in the case of developed countries and 24 per cent over then years in the case of developing countries, with minimum reductions per tariff line of 15 per cent and 10 per cent, respectively. The reductions in the tariffs of developed countries—which account for about two-thirds of world imports of agricultural products-indicate an average percentage reduction of 37 per cent. With respect to individual product categories, developed countries will cut tariffs by above-average amounts on oilseeds, flowers and plants; and cut tariffs by below-average amounts on sugar and dairy products, with other product categories close to the average cut. In the categories of "topical products", which account for half of exports of developing countries of agricultural products, a 43 per cent reduction in tariffs will be implemented by developed countries.

Current access opportunities will be maintained on terms at least equivalent to those existing prior to the tariffication process. However, for those products where tariffication took place and imports were less than 5 per cent of domestic consumption because of the existing restrictions, minimum market access commitments, implemented through tariff quotas on an MFN basis at a low or minimal tariff rate, are required. Figures on the increased market access in terms of tonnage resulting from minimum access commitments indicate that substantial increases in market access occur for coarse grains (1,757,000) and rice (1,076,000), as well as for other products. With regard to commitments on export competition, the quantities of exports which can be legally subsidized must be reduced by 21 per cent. Furthermore, total export subsidy outlays will decline by 36 per cent, from $21.3 billion to $13.7 billion by the end of the transition period. The importance of this commitment is that, on average, developed countries subsidized annually during 1986-90 48.2 million tons of wheat, 19.5 million tons of coarse grains, 1.8 million tons of sugar, 1.2 million tons of beef, etc. With regard to commitments on domestic support to agricultural producers, total outlays (in terms of the Aggregate Measurement of support) will be reduced by 18 per cent, from $197, billion to s162 billion by the end of the transition period.

The new market access opportunities for agricultural products which will result from the Uruguay Round—as a result of a change in border measures, and policies relating to export competition and domestic support—will be of particular interest to developing countries exporting temperature food products. More generally, multilateral discipliners on trade-distorting practices in agriculture are expected to stabilize world food markets in the coming decades, providing potential trade opportunities for developing countries and reducing fluctuations in food import bills, However, the potential situation in net food-importing developing countries is of particular concern. Potential problems relating to least-developed and net food importing developing countries are the subject of the Decision on *Measures Concerning the Possible Negative Effects of the Reform Programme on Least-Developed and Net Food Importing Developing Countries.* The Decision sets out objectives with regard to the provision of food aid, the provision of basic foodstuffs in full grant form and aid for agricultural development. It also refers to the possibility of assistance from the International Monetary Fund and the World Bank with respect to the short-term financing of food imports. The WTO Committee of Agriculture will monitor the implementation of the Decision.

WTO Agreements Covering Trade in Goods

GATT 1994: The cornerstone of trade relations in the area of goods. Differential and more favourable treatment to developing countries and to least-developed countries is permitted under the 1979 Enabling Clause with respect to tariffs in the context of the Generalized System of Preferences (GSP) and non-tariff measures, notwithstanding the most-favoured-nation clause, and with respect to regional or global arrangements concluded by developing countries.

Agreements integrating practices otherwise on the margin of GATT rules: Includes trade-related investment measures (TRIMS) (which can be found to be inconsistent with the national treatment provision or the prohibition or quantitative restrictions), such as local content requirements or trade-balancing requirements. GATT inconsistent TRIMs are required to be notified and eliminated within a transition period of two years (developed countries), five years (developing countries) or seven years (least-developed countries). A further extension may be requested by developing and least-developed countries. The Agreement on Safeguards prohibits the use of "grey-area measures", such as voluntary restraints or orderly marketing arrangements; such measures are to be notified and eliminated.

Agreement on Textiles and Clothing: Provides for the eventual elimination of the Multi-Fibre Arrangement (MFA) after a ten-year transition period. In place since 1973, the MFA currently groups eight "importers"; of these, Austria, Canada, the European Communities, Finland, Norway and the United States apply restrictions under the MFA, while Japan and Switzerland do not. The other participants in the MFA are the "exporters" (mainly developing countries), whose exports or part of their exports covered by the MFA are subject to bilaterally agreed quantitative restraints or unilaterally imposed restraints on imports, typically applied at the product level but in some cases to various aggregates as wells.

Trade in Services

The General Agreement on Trade in Services (GATS) is the first multilateral agreement on trade that has its objective the progressive liberalisation of trade in services. It provide for secure and more open market in services in a similar manner as the GATT has done for trade in goods. The Agreement covers trade in all service sectors and the supply of service in all forms.

The GATS has two components: the framework agreement containing 29 Articles and a number of Annexes, Ministerial Decisions etc., as well as the schedules of commitments undertaken by each Member to bind the existing degree of openness or remove existing resurrections.

Of importance to developing countries is the fact that virtually all Member have made commitments on the movement of natural persons, even if these are circumscribed by the requirement of intra-corporate transferee status. In addition, commitments made by developed countries generally cover the cross-border supply of labour-intensive services such as computer-related services, professional and construction services. Further, most developing countries have committed themselves to bind or liberalize tourism and travel service, including, for example, the liberalisation of foreign investment restrictions for hotel and resort operators. These commitments are likely to improve the supply capacity of this key sector, which provides the major source of foreign exchange earnings in a number of island developing countries and least-developed countries. In addition, a number of developing countries have taken the opportunity the GATS provides to schedule commitments, thereby binding their own domestic reform process. Improvements in the quality of service that will result from liberalisation and increased competition will contribute to improved efficiency, consumer welfare and growth in developing countries as well as all other countries.

Intellectual Property Rights

Under the WTO, the number of countries providing intellectual property protection will increase over time. Developed countries have one year to meet their obligations, developing countries have five years and least developed countries have eleven years, with the possibility of an extension. Special transitional arrangements apply in the situation where a developing country does not presently provide patent protection in a particular chemical.

Adherence to the Paris and Berne Conventions is fairly widespread among developing countries. Many developing countries already provide minimum standards of intellectual property protection on a national treatment basis, although the scope of such protection varies significantly. Potential benefits for developing countries emerging from the Uruguay Round include a framework more conducive to domestic research efforts and to technology transfer and foreign direct investment. There will, however, be additional administrative burdens of enforcing such rights (specifically dealt with under the TRIPS Agreement), potentially higher royalty payments and adjustment costs for industries which, in the absence of domestic legislation in the areas, were producing goods that would be considered as counterfeit in the future. These will also be requirements relating to patents which may well mean an increase in prices of certain goods in some developing countries. Pharmaceutical and agricultural products present examples. These increases are expected to be small, and there are provisions in the Agreement itself to minimize any adverse implications for developing countries.

Dispute Settlement

From the perspective of developing countries, it should be noted that the elements of the 1966 Decision on Dispute Settlement will continue to apply under the WTO dispute settlement procedures. Although this Decision has seldom been used, mainly because developing countries have only recently become more frequent users of the GATT dispute settlement procedures, it contains features of specific interest to developing countries, including automatic access to the "good offices" of the Director-General of the GATT/WTO to mediate and seek to find a satisfactory resolution to the dispute, and shorter time-limits in which panels must complete their deliberations.

Monitoring of Trade Policies

The TPRM provide for a Trade Policies Review Body to examine regularly the trade policies and practices of Members, every two years for the four major traders (the EU, US, Japan and Canada), every four years for the next sixteen leading traders, and every six years for the remaining traders, although longer intervals may be prescribed for least-developed countries.

The TPR process has helped countries assess their trade and economic reforms, and may have contributed to some portion of the liberalisation that has taken place under the Uruguay Round. In the future, the TPR process will help WTO Members evaluate their implementation of the Agreements, as well a provide an early warning of trends of potential concern to all participants in the trading system.

28

The Dematerialisation of the World Economy

The first Industrial Revolution marked the transition from robber – and - plunder colonialism to the systematic development of the "overseas" territories in the framework of the international division of labour between raw materials suppliers and manufacturers of finished goods. There was an "historic integration" of the colonised areas in the development of their parent—states. What will the third Industrial Revolution do for the Third World ? Will it now come to an "historic separation" ?

The end of the East-West conflict was reason enough to talk about a radical change in world politics. But at the same time an upheaval in the world economy is taking place that possibly will have even wider impacts. As a reference point for the following thoughts, three dimensions of this change are pointed out:

1 The upgrading of processing information rather than materials as object of economic activity (technological dimension) ;

2 the evolvement of global communications networks (sociocultural dimension) ;

3 the change of the nature of work (socio-economic dimension).

All three dimensions can be summarised under the buzzphrase "tertialisation of the world economy."

In that respect, talk of the "Third Industrial Revolution" is misleading. It is not about a third epoch of industrialisation, but about the beginning of a de–industrialisation, the transition from the industrial to the information society.

Historic Separation?

In the 1960s and early 1970s, there was often talk of the Third World as the Third Sector of the world economy. Also then the Third World was not much more than an "imaginary community". But as such it had a certain significance in world politics. This implied not only its strategic role in the East – West conflict and its ideological function as the supporter of different "third paths" between capitalism and socialism. It was also about the Third World's attested "chaos power". That linked the fear (in the North) and the hope (in the South) that the developing countries would be in a position to cut off the industrial nations from supplies of important raw materials, thus putting them under pressure. But it was soon seen that both sides had over estimated this possibility, even with regard to oil. Instead of supply bottlenecks arising, raw materials prices plummeted. For some commodities, the fall in prices exceeded those of the Great Depression of 1929/30.

This was due, inter alia, to the conjunction of lower demand from the industrial nations and expansion of production by the raw materials suppliers. Business activities dependent upon the supply of raw materials are tending to lose importance compared with the overall development of the global economy. The reason for this is to be seen in the transition from a material to an information economy.

This transition is taking place in line with the revolutionizing of data transmission and the expansion of financial transactions which are not directly related to changes in the production of materials. The speed of the changes is remarkable.

However, the dematerialisation of business activities does not lead to decoupling of the Third World from the world economy. Declining market shares in world trade are not the expression of separation, but a loss of the affected countries positions in the world economy. Thus, the impact of dematerialisation is "only" that the negotiating positions of raw materials suppliers vis–a–vis the industrial nations will deteriorate further.

Differentiation of the Third World

But the radical change in the global economy is affecting some developing countries worse than others. Sub–saharan Africa, and some countries in West and South Asia and Latin America are being pushed back further. The oil–producing countries with their high per capita export earnings will be able to hold their positions in the world economy for some time to come. The threshold countries of East and South East Asia can expand theirs so long as they can continue to attract a growing share of global industrial production, and at the same time participate in the tertialisation of the world economy in the shape of rapidly-growing financial transactions. Thereby it should be noted that the degree of tertialisation in itself is not an adequate indicator for economic avant-gardism. Brazil exhibits a high degree of tertialisation in combination with a low macroeconomic development dynamics. A good part of its tertialisation is being achieved by speculative financial transactions with their inherently greater risks and uncertainties than in the industrial countries. Such dangers have been demonstrated by Mexico's peso crisis and its repercussions on the whole of Latin America.

In some Third World countries, a "location annuity" has replaced the old raw materials one. Here it's about providing locations for off-shore transactions which offer international capital traders a maximum of freedom of movement combined with low taxation. Suitable for such operations are small countries which, despite low levy rates, achieve significant income in macroeconomic terms.

The radical changes in the world economy are spurring the differentiation of the Third World Without, however, necessarily fostering a dissolution of the Third World as an "imaginary community". It is precisely the advanced countries of East and South-East Asia that are showing a certain interest in the formulation of joint positions of the "South" in order to secure their own positional gains in the global economy. It's not by chance that the non-aligned countries and the Group of 77 have formed a joint coordination committee, and that the ASEAN countries are changing course on the international human rights policy.

Hitherto, the developing countries' strategy was to broaden the concept of human rights as a justification for demands on the industrial nations. But of late some developing countries, led by the ASEAN states, have questioned the universal validity of human rights even after their universality was confirmed by consensus at the Conference on Human Rights in Vienna in 1993. Playing a role in this policy is the governments' fear that due to the expansion of global communications networks, the behaviour patterns and preferences of their own people could in some way become similar to those of the West. As the rulers see it, that would be detrimental to the continuation of the development models practised so far.

Internet Creates New Cultural Dimension

Much information which Asian governments view as subversive in already globally available on the Internet. The old struggle over the world information order, which at first was primarily a clinch between East and West, is thus taking on a new dimension. For with the growing importance of computer literacy to a country's ability to assert itself on world markets, the Asian threshold countries have not only an interest in controlling the on-line communication but also to expand it and the know-how that it requires.

Even the critics of any interventions in the internet and other global communications networks must admit that modern communications technologies are politically blind and their use in itself does not represent progress. The setting up and expansion of global information highways will offer forum not only to people who want to use it for education and enlightenment, but also to all shades of fundamentalists. These highways will not necessarily bring the misery of many Third World regions closer to the industrial countries, but possibly rather strengthen the tendency to process all world events as entertainment.

Global Two-thirds Society

The gravest aswpect of the current upheaval in the world economy is its negative impact on jobs. The information economy needs for fewer workers than an economy based on materials. Instead, the demands on the skills of the workers are growing. Twenty per cent of the world workforce will in future be employed as (overworked) "intelligence workers". Eighty per cent will work part–time, if they are not underemployed or jobless. So the tertialisation of the global economy delivers more underemployment rather than more leisure time. The workers who are rationalised out of their jobs in the industrial sector cannot be absorbed by the service sector because it, too, is not left untouched by rationalisation measures. The civil service is also cutting back on staff. At all levels, there's a race to make the greatest possible savings on payrolls. At the same time, there's growing pressure to cut costs in providing for the victims of this development. That means thinning out the social security safety net.

The bottom line is that the two-thirds society, which developmental action groups hitherto assumed was limited to the Third World, is spreading worldwide. That, however, will not in the foreseeable future lead to an amendment of the North-South disparities. It's true that the change in the global economy is taking place faster, and to a greater extent in the industrial nations. But rationalisation is also happening in the developing countries in a bid to boost their competitiveness. So the upheaval in the world economy aggravates the problems which exist in a majority of the developing countries, while creating new ones in the industrial nations. The need for action on the North-South policy is growing, while the industrial nations. The need for action on the North-South Policy is growing, while the industrial nations' scope for concessions and compromises is shrinking. The new social question which is now crystallizing at global level is not being answered. The consequences are unforeseeable.

Another Loser ?

It's more probable that a sharpening of the North–South confrontation is to be reckoned with. For the industrial nations will attempt to keep the social costs of the information economy at bay for as long as possible. The trade unions will thereby compete with the developing countries for jobs for their memebers. But this policy has its limits precisely because of the peaking of the problems in the industrial nations. Overstepping these limits means war, and passively accepting them chaos and social decay. Solutions could be sought in two directions: effective taxation of the information economies, and the creation of jobs in the non-profit sector. But it's possible there are no global solutions for global problems. That would mean for at least part of the Third World a renewal of the old debate on partial decoupling from the world economy.

29

Crisis and New Orientation of Development Policy

The poverty in the South, the dislocations in the East, and the orientation crisis in the North are not isolated phenomena. Rather, they represent an alarming amalgamation of dangers that are globally interlinked.

The low effectiveness of international economic and development policy is rooted in two outdated paradigms on which the present worldwide strategy of economic development is based, namely that:

1. The Western social and economic model optimizes the activation of productive forces—independent of the development stage of a country and its culture and therefore is best suited to satisfy basic needs.
2. It is possible to launch the development of a society from the outside within a few decades-without regard to its cultural and historical background—through external input of money, goods, technology, expertise, and personnel.

The twin paradigms of the timelessness and transferability combined with cultural ecological, and financial restrictions—have led international cooperation and development down the wrong path.

Only if we acknowledge the true dimensions of the global dangers, if we recognise the limitations and shortcomings of existing political instruments, and identify outdated theories and contradictory special interests, can we outline the cornerstones of a new policy of global cooperation.

Cornerstones of a New Development Policy

Starting with critical review of the shortcomings and paradigms of the prevailing development strategy, the following ten cornerstones of a new development policy are offered for discussion:

1. *Broaden the Concept of Development*

Whether a society is considered developed depends on the size of its percapita Gross National Product (GNP). Accordingly, the world is divided into a developed, semi-developed, and underdeveloped world. The yardstick for development, which has become the norm in the industrial countries, is one-dimensional: It only measures the monetary value of goods and services that are exchanged in the marketplace. This standard is too narrow economically because, it compresses the multitude and complexity of cultural, societal historical, social, and human values into a single economic category.

At the most, there can and should be agreement on what development and process should not bring about: Inability to find enough work to meet the most basic needs; exploitation and oppression of people; loss of cultural wealth and institutions; destruction of natural resources. These, however, are the very values that are scarified by the prevailing development strategy. In the future, development policy must do all it can to stop the loss of skills and self-reliance, the plunder of natural resources, the erosion of cultural values, the violation of human dignity and human rights. Initiatives must prevail which are orientated on these values, and not just on the GNP.

2. *Concentrate Development Strategy on the Internal Potential of Developing Countries*

There must be an end to the manic fixation of development strategy on external inputs and external markets. A new development policy must, above all, improve internal conditions for a productive economy, promote domestic production factors on a broad basis, protect cultural and natural resources, and greatly increase the domestic supply of basic goods. Wherever external inputs are unavoidable, credits must be strictly tied to the productivity and the ability of a country to absorb transfers. External transfers should be concentrated on "Software" for health, education, and social participation, administrative, and legal jurisdiction. Such an approach could also promote training and indigenous technologies, which are so important for economic development.

The set-up and expansion of the productive sectors must be decided, planned, and implemented by the developing countries themselves, and they must assume full responsibility. The external pressures, which force the developing countries into full integration with the world market, must be removed. This presupposes a structural reduction of interest rates.

3. *Make Development Policy a Central Feature of Polities*

Development policy must take the lead in mobilizing the various political forces and government departments to join the fight against the growing global dangers. It must ensure that the actions of all political departments are compatible with development policy is possible only if it becomes the central task of all political sectors, comparable to social and environmental policies, and the central goal of all policies. If development policy is to become a central task, development problems must become a priority in parliament and government. Society must understand that it is in the national interest to accept great global responsibilities.

4. *Reform the World Economy*

The industrial countries must abolish their protectionism in agriculture as well the processed goods sector. Simultaneously, the developing countries need to be protected selectively and for a limited time against imports from the industrial countries. The undifferentiated structural adjustment policies imposed by the IMF must be revised. The trend toward regionalisation of the world economy should not be opposed; rather, in the interest of both South and East, it must be regulated constructively to form a new, regionally based world trade structure.

A reform of the international finance system is urgently needed: Interest and exchange rates should not mirror the national interests of the big industrial states and the special interests of large banks and venture capital. Rather, they must reflect the global interest in monetary stability lower and stable interest rates, and sufficient development financing.

However, strengthening the international financial institutions is in the global interest only if the countries of the southern and eastern hemispheres are allowed to exert some influence. An international financial court must guarantee that violations of strict regulations to ensure international stability and solvency can be protested in a court of law.

5. *Redesign the Industrial Society*

As a global social and environmental policy, the new development policy must induce the industrial countries to give up their excessive consumption of air, water, soil, resources, and space. Increased utilisation of energy-conservation measures and environmentally friendly technologies is overdue. The economic and social policies of the industrial nations must promote balance rather than growth. This requires radical changes in traditional economic thinking, habits, structures and processes.

In view of limited world resources, unsatisfied existential needs in South and East, and continuous population growth in the south, the only premise for the future can be: Growth rates in the South must be higher than in the North, but they should no longer be in the North, but they should no longer be induced primarily by growth in the North. If economic policies continue to call for the North to provide the locomotive, the North will have to continue to acquire more resources than the south.

The North must relinquish the remaining growth frontiers to the South and East. The South must use this opportunity to activate its internal dynamic potential rather than integrate its economy with the North. However, ecological and social controls must be established at a much earlier stage than was the case in Europe.

6. *Strengthen Development Cooperation*

The share of official development assistance as a percentage of GNP, which dropped from 0.48 per cent in 1982 to 0.34 per cent in 1995 must be gradually raised again and reach at least 0.7 per cent in the year 2000—a goal which OECD established as early as two decades ago and which was reconfirmed at the Rio Earth Summit.

However, we must not succumb to the illusion that a doubling of ODA funds will even remotely meet the financial needs of South and East. State development policy must use it scarce public funds more effectively in the future. It must use restraint whenever partners in the developing countries can accomplish a task on their own and private initiatives and private enterprise are more competent to do the job. The government should be directly engaged only when it can be relatively more productive. Otherwise, it should limit itself to subsidizing private organisations.

7. *Now Orientation for Development Cooperation*

The state and its implementation agencies must abandon all direct responsibility for any projects which require unbureaucratic action, economic efficiency, and long term productivity. It must make a much greater effort to involve NGO's and private venture capital in development projects. At the same time, the state must insist and guarantee that private actions are compatible with social and ecological concerns.

In the future, the main thrust of government projects should be the promotion of the internal potential of a country. This comprises the political and administrative framework conditions of a humane, socially and ecologically sound development: Constitutional government, social institutions which facilitate broad participation of the population in politics, society, and economy; efficient savings, credit fiscal and financial systems; mechanisms for income, property, and land distribution which, promote productivity, justice, and social peace. In additional of this "software" of development, the following is needed: A regimen for the protection of resources and environment; measures to prevent the short term sellout of natural resources; elementary and general education and training, health care and social safety nets; capacities to develop science and technology.

8. *Reduce the Debt Service and Activate Private Capital*

Public funds must be used be to a greater degree for the financial rehabilitation of highly indebted countries in South and

East; external demands for interest and principal payments must be adapted to the economic capacity of the respective country and its ability to execute external capital transfers.

Within the framework of international insolvency regulations, initiatives must be developed as condition for the continuance of the present rules for write-offs—which ensure effective cooperation from the banks and alleviate the heavy burden of private credits, with their high interest rates.

State development policy and private business interests should supplement each other. Government promotion of private enterprise initiatives for exports, investment, and employment in the developing countries must take into account their compatibility with development. In reverse, private engagements that effectively promote development must be actively supported by the government. A separate line item must be established in the development budget for such activation of private capital

9. *Set Regional Priorities*

State development cooperation has been scattering its scarce funds that not only among too many sectors, but also among too many partners. In the future, public funds must be concentrated regionally. More emphasis must be placed on regional programmes, and development cooperation with threshold countries must be enhanced. A portion of public funds should be set aside to provide an incentive for be set aside to provide an incentive for threshold countries to assist the poorer nations in their own region as well as deal with poverty in their own country.

The new development policy could then also help lessen ethnic-national conflicts and promote peace by sponsoring regional cooperation in joint development projects. For this purpose, regional development funds must be set up for cooperation in the transportation, energy, trade, and finance sectors and last, but not least for regional security systems and disarmament. Such regional funds could also provide the means to project refugees and improve their prospects for an eventual return to their homelands.

30

A New World Order for Whom?

"Four holocausts" humanity has produced are breeding the seeds of our own destruction: war and militarisation, human oppression, economic destitution and environmental destruction. The "new world orders" on offer can satisfy only the minority of the world's rich and will ultimately only exacerbate these four trends towards global annihilation. But there is also hope in the "grassroots world order" leading to the civil society, democratisation and social mobilisation as the only way for the planet to survive.

As we approach the end of the 20th century and a new millenium, humanity is faced with four conditions of its own making, so serious in terms of their present destruction of life and risks for the future that they warrant a description as "four holocausts".

The first holocaust is that of war and militarisation. After the Gulf War it is clear that, far from preparing the way for world peace, that conflict has unleashed a new global arms race in the weapons whose brutal effectiveness was so clearly demonstrated in Iraq. The second holocaust is that of human oppression, the violent denial by governments of the basic personal, civil, political and economic rights of their citizens, which routinely persist in a majority of countries of the world. The third holocaust is that of economic destitution, the mass poverty of a fifth of the world's population, leading to endemic malnutrition, disease and death, not least among children. And the fourth holocaust is that of environmental destruction, which is gradually or not so gradually rendering the planet uninhabitable, as witnessed by the growing millions of environmental refugees whose bankrupt ecosystems can no longer support them.

None of these circumstances is entirely new to our age, of course, War, repression, destitution and ecological degradation have often been part of the human condition. What is new about the current situation is its global nature. All humanity, the whole earth, is now at risk. And it is all humanity which must be party to a response, if it is to be successful.

Against this threatening background, several new developments stand out. First, is the collapse of communism, taking out of the bounds of credibility socialism's millenarian dream of abolishing the market and inevitably replacing capitalism through the onward march of history. Second, there are the twin trends of globalisation and interdependence: globalisation especially of the economy, interdependence especially through environmental impacts. This is the context for any discu[illegible] f a "new world order".

ıne New World Orders

There are, broadly, three kinds of new world orders currently on offer. The real world is almost certainly going to be mixture of all three, but the balance between them will be crucial in deciding whether we will successfully manage to face, and overcome, the holocausts raging among us.

The first kind of new world order may be called the "neoliberal". Its most important component is the untrammelled operation of what we would call the global "free market". At once we must qualify this terminology by noting that the freedom bestowed on someone by the market is in direct proportion to the amount of property owned by that person within it. In a free ma[illegible] hose who own the means of production are free to [illegible]e what, when and where they want, and largely to determine the conditions of production. Those who own the means of consumption can similarly scour the world for products to satisfy their wants. The extent to which this "freedom" is far from universal is shown by the fact that, within regard to consumption some 23 per cent of the world's population control

85 per cent of the income which is consumptions prerequisite and ownership of the means of production is more concentrated still. The neoliberal world order thus represents a good deal for perhaps a quarter of the world's population, but has precious little to offer the rest.

A Global Order Framed by International Institutions

The second kind of new world order may be termed the "social democratic" and is roughtly that advocated in the 1970's by the proponents of a New International Economic Order. These two new world orders clearly different but they also share several characteristics which seem to be more significant than their differences. First, they are explicitly western-oriented and homogenizing. They view the world through the eyes of western science and western culture simultaneously devaluating the knowledge and accumulated wisdom of the great majority of human kind. The paradigm society, towards which all others are supposed to be developing or aspiring to develop, is that of the United States.

Second, both these world orders are economistic. Human progress and development to them means economic developments, still usually measured by the level and growth of GNP per person. No social or cultural tradition or aspiration is allowed to stand in the way of this "development". Third, both world views envisage top-down decision-making for administration and control. For the neoliberals the dominant influence is exercised by the owners and managers of transnational capital. For the social democrats their influence is balanced by the interventions of international and national bureaucrats. Neither world order places great store on consultation with, let alone decision-making by, ordinary people in their communities.

In contrast to these two new world orders, it is possible to posit a third, here called the grassroots new world order, which takes as is principal focus neither the market nor the state (national or international), but civil society, the networks of family, community and voluntary association acting for social reproduction, reconstruction or reform. This new world order has characteristics diametrically opposed to those shared by the first two discussed. Its impulse derives explicity from the bottom up, drawing on the capability and creativity of those united by shared values and interests, at the local level or in wider networks. Their world-view is one of cultural diversity, of one world comprised of many different villages rather than a homogeneous global village modelled on the US. They perceive human development to be holistic, with the economic dimension integrated with, or embedded in, a broader social, ethical and ecological reality. And they proceed from an ethical basis that strives for ecological sustainability, social justice in distribution and broad participation is cultural, political and economic life.

Looking to the Grassroots World Order

After this thumbnail sketch of these three views of different dominant global processes, one can ask which of them or, more realistically, what balance between them, will be best able to put an end to the four holocausts tormenting humanity. There is only one convincing answer: the dominant thrust must be towards the grassroots new world order. There are several reasons for this. Most obviously, it is indisputable that it is the forces of the market and the state that have not only failed to douse, but have actually fanned the flames of all four holocausts. It is states that go to war and waste the commonwealth on weaponry. It is states that are responsible for the great majority of violence and repression against ordinary people. It is states, often supported by multilateral governmental organisations such as the World Bank and IMF, that have the so called Third World intervened massively in the subsistence, largely nonmarket economies of the people, and redistributed their resources, redefining their very rights to property, in favour of industriatlisation and market exchange. But these enhanced markets have then spectacularly failed to provide alternative subsistence for those dispossessed, leaving them by the million impoverished, marginalized and destitute. Moreover, this process has set in train two great engines of environmental destruction: industrialisation itself, with its toxic pollution, social èrosion, water and ozone depletion and climate destabilisation; and the depredations of the rural dispossessed, forced into forests or onto marginal lands, deforesting, making deserts, extinguishing species and multiplying in numbers of their desperate efforts to stay alive.

Civil Society Mobilizing for Human Survival

In contrast, it is civil society that has mobilized explicitly against the four holocausts. The great social movements of our time are those for peace and human rights, for justice and development, and for environmental conservation. It is independent, non-violent associations of civil society that have sought explicitly to address these issues, meeting at best indifference from organisations of the market and the state, at worst outright hostility.

This is not at all to say that the market and state are irredeemable, or that they have no role in combating the four holocausts of destruction. On the contrary, they each have a vital contribution to make but, it order to make it, each must first be transformed. The market failures caused by great concentrations of wealth and power, ubiquitous externalities, the tyranny of small decisions and positional goods can only be resolved by determined and democratic state action. But most governments are not democratic. On the contrary most are unrepresentative and self-serving, many are vicious and corrupt. And the only force that can democratize them is that of civil mobilisation and organisation.

Viewing the immense power of today's concentrations of wealth, and the remoteness of most governments from their people, one may feel despair at the prospect of civil society being able to harness these forces to the common good. And indeed there is no certainty that they will be thus harnessed. The four holocausts may simply run their awful course. But since 1989, at least, there is proof positive in the peaceful revolutions in Eastern Europe and the Soviet Union, that civil society can overturn seemingly omnipotent despotic structures. And all over the world the various movements for peace, justice and the environment are organizing for human survival. The stakes have never been so high and the outcome is uncertain; but there are legitimate, and inspiring, grounds for hope.

Bibliography

De Soto, H., 1990. *The Other Path; The Invisible Revolution in the Third World,* Reprint edition. New York: HarperCollins.

Doha Development Agenda, 2001. The Ministerial Declaration and other Decisions and Declarations from the Doha Ministerial Conference, Available: http://www. wto, org/english/tratop_e/dda_e/dda_e.htm.

English, P., B. Hoekman, and A. Mattoo, 2002, *Development, Trade and the WTO:A Handbook*. The World Bank, Washington D.C.

Feketekuty, G., 1988. *International Trade in Services: an Overview and Blueprint for Negotiations,* Cambridge, MA: American Enterprise Institute/Ballinger.

Finger, J.M., 1993. *Antidumping: How It Works And Who Gets Hurt.* Ann Arbor: Univ. of Michigan Press.

Finger, J.M., 2001. "Implementing the Uruguay Round Agreements: Problems for Development Countries." *The World Economy* 24(9, September): 1097-108.

Finger, J.M. and J.J Nogues, 2001. The Unbalanced Uruguay Round Outcome: The New Areas in Future WTO Negotiations. *Policy Research Working Paper No. 2732,* The World Bank, Washington, D.C.

Finger, J.M. and L. Schuknecht, 2001. "Market Access Advances and Retreats: The Uruguay Round and Beyond." In B. Hoekman and W. Martin, eds., *Developing Countries and the WTO: A Pro-active Agenda*. Oxford: UK and Malden. Also available as Policy Research, Working Paper No. 2232 at http://www.worldbank.org/research/trade.

Finger, J.M. and P. Schuler, 2000. "Implementation of Uruguay Round Commitments: The Development Challenge." *The World Economy* 23(4, April): 511-25. Also available as Policy Research Working Paper No. 2215 at http://www.worldbank.org/research/trade.

Finger, J.M. and L.A. Winters, 2002. "Reciprotocity." In P. English, B. Hoekman, and A. Mattoo, *Development, Trade and the WTO: A Handbook*. The World Bank, Washington D.C.

Finger. J.M., M.D. Ingco, and U. Reincke, 1996. *The Uruguay Round: Statistics on Tariif Concessions Given and Received*. The World Bank, Washington, D.C.

Finger, J.M., F. Ng, and S. Wangehuk, 2001. Antidumping as Safeguard Policy. Policy Research Working Paper No. 2730, The World Bank, Washington, D.C.

Francois, J.F.,B. Mc Donald, and H. Nordstrom , 1996. "The Uruguay Round: A Numerically Based Qualitative Assessment." In W. Martin and L.A. Winters, eds., *The Uruguay Round and the Developing Countries*. Cambridge: Cambridge: University Press.

Harrison, G.W., T. F. Rutherford, and D.G. Tarr, 1996. "Quantifying the Uruguay Round." In W. Martin and L.A. Winters, eds., *The Uruguay Round and the Developing Countries*. Cambridge: Cambridge University Press.

Hudec, R.E., 1970. "The GATT Legal System: A Diplomat's Jurisprudence." *Journal of World Trade Law* 4:615-65.

International Intellectual Properly Alliance (IIPA), 2002a. "Description of the IIPA." Available: http://www.iipa.com/aboutiipa .html.

International Intellectual Properly Alliance (IIPA), 2002b. "Statistic."Available: http://www.iipa. com/statistics. html.

Martin,W. and I,. A. Winters, 1996. *The Uruguay Round and the Developing Countries*. Cambridge; Cambridge University Press.

Martin. W. and L.A. Winters, 1996. "The Uruguay Round: a Milestone fo the Developing Countries." In W. Martin and L.A. Winters, eds., *The Uruguay Round and the Developing Countries*. Cambridge: Cambridge University Press.

Maskus, K.E., 2000. *Intellectual Property Rights in the Global Economy*. Institute for International Economics, Washington, D.C.

Michalopoulos, C., 1999. "The Developing Countries in the WTO." *The World Economy* 22(1) January.

O'Neill, T. and G. Hymel (contributor), 1995, *All Politics Is Local: And Other Rules of the Game*. Reprint edition. Massachusetts: Adams Media Corporation.

Panagariya, A., forthcoming. "Developing Countries at Doha: A Political Economy Analysis." *The World Economy*.

Petersen, M. and D.G. McNeil J., 2001. "Maker Yielding Patent in Africa for AIDS Drug" *The New York Times*. 15 March. Page 1.

Preeg, E.H., 1995. *Traders in a Brave New World*. Chicago and London: University of Chicago Press.

Reichman, J.H., 1998. "Securing Compliance with the TRIPS Agreement after US v India." *Journal of International Economic Law* 1(4, December): 603-06.

Ricupero, R., 2000. "A Development Round: Converting Rhetoric Into Substance." Paper presented at the Symposium on Efficiency, Equity and Legitimacy: the Multilateral Trading System at the Millennium, 1-2 June, John F. Kennedy School of Government, Harvard University, Cambridge, Massachusetts.

Shaffer, G., 2002. "The Law-in Action of International Trade Litigation: The Blurring of the Public and the Private." University of Wisconsin Law School, Madison Manuscript.

Winham, G., 1986. *International Trade and The Tokyo Round of Negotiations*. Princeton: Princeton University Press.

Winters, L.A., 2002. "Doha and the World Poverty Target." Paper prepared for the Annual Bank Conference on Development Economics (ABCDE), 29-30 April, World Bank, Washington, D.C.

World Bank, 2002, *Global Economic Prospects and the Developing Countries*. The World Bank, Washington D.C.

World Trade Organisation (WTO), 2002a. "WTO Secretariat Budget for 2002." Available: http://www.wto.org/english/thewto_e/budget_e.htm.

World Trade Organisation (WTO), 2002b. Pledging Conference to provide sound financial basis for Doha Agenda. Available: http://www.wto.org./english/news_e/pres02_e/pr277_e.htm.

Zeller, T.W., 1992. *American Trade and Power in the 1960s*. New York: Columbia.

Index

D

E